Praise for [illegible]*ons*

Sentience, cognition and intelligence are emerging as [illegible] faculties of all life which has evolved on the Earth. Most of these living systems are much older than humanity and obviously are well integrated to support life. In *Biocivilisations*, Predrag Slijepčević makes clear that the sentient life is essential for the habitability of our planet and that humans should step down from the so-called crown of evolution model in order to appreciate our true position within the complex network of life. Only then will our civilization improve its rather doomed prospects for survival.

Dr František Baluška, Institute of Cellular and Molecular Biology, University of Bonn

Read this book if you would like to understand the intelligence of living systems. Civilisation did not just start with *Homo sapiens*. Life cannot be reduced to pure mechanism.

Dr Denis Noble, Emeritus Professor of Cardiovascular Physiology, University of Oxford; Fellow of the Royal Society; 2022 Lomonosov Grand Gold Medal laureate

A prodigious synthesis and a great, ambitious and informative book dovetailing multiple fields in its effort – largely successful I think – to light a match – and then blow on the fires of the coming 'Copernican biological revolution.'

Dorion Sagan

In *Biocivilisations*, Predrag Slijepčević tells stories about animals that create art, insects that do battlefield surgery, trees that perform scientific research, bacteria that create intelligent networks, and whole ecosystems that are organized with an efficiency that surpasses any human supply chain. Maybe you thought humans were the crown of creation. Maybe we humans have to learn humility and respect for the biosphere that birthed us. Maybe our future depends on it.

Josh Mitteldorf, PhD, coauthor of *Cracking the Aging Code*

Predrag Slijepčević's *Biocivilisations* brings together crucial developments in biological systems thinking – such as symbiogenesis, epigenetics, biosemiotics, Gaia theory and autopoiesis – under a comprehensive vision founded on the cosmological longevity and cognitive acumen of the bacterial microcosm and its planetary offspring: multicellular life in all of its forms and alliances. I highly recommend Slijepčević's *Biocivilisations* for those who would like to get effectively up to speed on the most cogent contemporary challenges to the physicalist-mechanistic techno-scientific mainstream.

Bruce Clarke, Paul Whitfield Horn Distinguished Professor
of Literature and Science, Texas Tech University,
Baruch S. Blumberg NASA/Library of Congress Chair in Astrobiology

Predrag Slijepčević's *Biocivilisations* is an unusually thought-provoking and ambitious book. It challenges the reader to abandon several centuries of assumptions about how to describe the living world in purely physical and mechanistic terms, a world governed by an evolutionary process that places human beings at the apex. Instead, the author sets out to 'explore the science of life in an entirely new way, through the concept of biology as a civilising force'.

It is important to recognize that *Biocivilisations* is more than just a new technical manner of describing life activities. The book carries a moral alert that heightens its scientific descriptions. In explaining the remarkable accomplishments of organisms that humans have all too often disparaged as lacking intelligence and technical capabilities, the author points out that all of these 'biocivilised' accomplishments have survived millions of years because their prehuman creators successfully integrated them into the ceaseless flow of the Gaian system. Modern human society, on the other hand, has produced many so-called 'advanced' technological creations which are unsustainable and threaten the Gaian system. Anyone who cares about our future on earth, therefore, has good reason to read this fascinating and highly original book.

Dr James A. Shapiro, Department of Biochemistry
and Molecular Biology at the University of Chicago;
author of *Evolution: A View from the 21st Century*

BIOCIVILISATIONS

A New Look at the Science of Life

PREDRAG B. SLIJEPČEVIĆ

Foreword by VANDANA SHIVA

Chelsea Green Publishing
White River Junction, Vermont
London, UK

Commissioning Editor: Jon Rae
Project Manager: Rebecca Springer
Developmental Editor: Brianne Goodspeed
Copy Editor: Clare Diston
Proofreader: Deborah Heimann
Indexer: Barbara Cuerden
Designer: Melissa Jacobson
Page Layout: Abrah Griggs

Printed in Canada.
First printing May 2023.
10 9 8 7 6 5 4 3 2 1   23 24 25 26 27

Our Commitment to Green Publishing
Chelsea Green sees publishing as a tool for cultural change and ecological stewardship. We strive to align our book manufacturing practices with our editorial mission and to reduce the impact of our business enterprise in the environment. We print our books using vegetable-based inks whenever possible. This book may cost slightly more because it was printed on paper from responsibly managed forests, and we hope you'll agree that it's worth it. *Biocivilisations* was printed on paper supplied by Marquis that is certified by the Forest Stewardship Council.

ISBN 978-1-64502-138-4 (paperback)
ISBN 978-1-64502-139-1 (ebook)
ISBN 978-1-64502-140-7 (audiobook)

Library of Congress Cataloging-in-Publication Data is available.

Chelsea Green Publishing
White River Junction, Vermont, USA
London, UK

www.chelseagreen.com

Contents

Foreword

Predrag Slijepčević's *Biocivilisations* is a vital book for our times. It provides the ontological and epistemic foundation for the transition we need to make as a species to sow the seeds of a liveable future.

The roots of various ecological crises, a global food crisis, and growing chronic health crises lie in a fallacy underpinned by anthropocentric arrogance: that humans are separate from nature and superior to other species, which have no intrinsic worth, value, intelligence or creativity. Rather than sentient beings, other species are reduced to objects that can be owned, manipulated for profit and pushed to extinction.

This worldview is incorrect and at the heart of our troubles. The truth is that *Homo sapiens* are not masters of the universe. We are members of one Earth family, diverse yet interconnected. This illusion of separateness and superiority over other species is a form of ecological apartheid. Apartheid means 'separation' in the Afrikaans language and ecological apartheid is an ideology that humans are separate from nature—as well as her conquerors, masters and owners. Anthropocentrism and ecological apartheid denies that humans are *part* of nature, not separate from her.

This worldview of separation engenders hierarchies and the illusion of superiority: of humans as superior to other species, men as superior to women, whites as superior to Blacks and coloured people, one faith as superior to the diversity of belief systems that have nourished cultural diversity from the beginning of history. Separation and superiority create structures of violence—violence against nature, violence against women, violence against every 'other', defined as lesser beings, with the objective of colonisation. Inequality and injustice are rooted in the false assumption of separation and superiority.

Anthropocentrism grows from mechanistic ontology and mechanistic science. The mechanistic world view has dominated human thinking over the last few hundred years. Central to the mechanistic world view is the ontological flaw that life is a machine and nature is made of inert immutable particles. A century ago, quantum theory showed this mechanical view to be false. The world is not constituted of static 'things', but of potential. Quantum theory also recognises interconnectedness through the principle of non-locality and non-separability, 'action at a distance'.

As physicist David Bohm suggests, 'the reason subatomic particles are able to remain in contact with one another regardless of the distance separating them is not because they are sending some sort of mysterious signal back and forth, but because their separateness is an illusion... At some deeper level of reality such particles are not individual entities, but are actually extensions of the same fundamental something.'

As Slijepčević reminds us in *Biocivilisations*, life is neither mechanism or thing. Rather, life is permanent change, and no organism is a machine. He calls mechanistic philosophy 'mechanophilia', or love for the machine, a symptom of what Gregory Bateson referred to as 'pathologies of epistemology'.

This pathological epistemology has eroded the ontology of a living intelligent Earth and living autopoietic organisms, as mechanistic reductionism was elevated as the only legitimate form of science. Francis Bacon, the so-called 'father of modern science', defined this shift from seeing nature as a nurturing mother and source of life to an object of economic exploitation, domination, and mastery as 'the Masculine Birth of Time'.

In *Masculus Partus Temporum* (*The Masculine Birth of Time*, 1603), a posthumously published text, Bacon writes, 'I am come in very truth leading to you Nature with all her children to bind her to your service and make her your slave.'

Since 'the dominion of man over nature rests only on knowledge', the key to man's mastery over the world lies in the epistemology of mechanistic reductionism. Science and knowledge were put into the service of dominating the Earth and its ecosystems (made up of, and reliant upon, communities of living organisms). In the epistemology of mechanistic reductionism, Earth and its non-human inhabitants have no rights, and there are no limits to ecological domination, exploitation and extraction. This is at the root of non-sustainability.

The illusion of separation from nature is at the heart of mechanistic thought, which treats nature as dead inert and passive. This gives license to exploit and violate nature. As I wrote in my book *Staying Alive*, 'If nature is dead, and the Earth is empty – Terra Nullius – then all violence against nature is defined as "human progress".' Technologies based on this mechanistic paradigm such as genetic engineering and Green Revolution agriculture tear apart the interconnectedness of the Earth at the level of the ecosystem, the cell and the genes—and have no method for even assessing the damage.

In *Biocivilisations*, Predrag Slijepčević brings us the principles of a *non-mechanistic* biology: 1) universal flux (*panta rhei*, everything flows); 2) agency (purpose and desire); 3) symbiosis (living together); and 4) mind (hyperthought).

The principles of biocivilisations are also the principles of just and sustainable societies. The extinction crisis offers an opportunity to transcend anthropocentrism and mechanistic science, as well as the separation that destroys ecosystems and human communities. The idea of the 'dominion of man over the universe' is a denial that we are members of one Earth family, rich in biodiversity—just as human beings the world over are one humanity, rich in cultural and biological diversity.

Mechanistic reductionism sowed the seeds of ecological apartheid and human apartheid, of anthropocentrism and capitalist patriarchy. Those who have profited from anthropocentrism continue to attempt to establish their 'dominion over the universe' through mechanistic ideology. Since Bacon's time, intelligence and knowledge have been treated as exclusive to privileged and powerful men. Nature's intelligence, women's intelligence, the intelligence of farmers and peasants, the intelligence and knowledge of indigenous peoples was denied. Now an even more extreme form of anthropocentrism and mechanistic thought is being proposed as the inevitable human future: transhumanism based on 'enhancement' of humans by machines and the rule of artificial intelligence (AI).

Our times call for a Gaian Democracy based on epistemic democracy and the democratization of intelligence. *Biocivilisations* shows us that all life, from the tiniest microbe to Gaia as a whole (a superorganism), is intelligent. 'Organisms/ agents are epistemic systems capable of understanding their surroundings', Slijepčević tells us. The intelligence of organisms is ecological. Widening our understanding of intelligence to include all life reveals the limitations of transhumanism. Democratising intelligence shows that rule by AI is not inevitable. Evolution can find a path forward through ecological intelligence.

As Slijepčević writes, 'Post-biological evolution is another prejudice cooked up in the kitchen of a mechanistic interpretation of life.' Organising networks and communities of relationship is also not unique to humans. Biocivilisations of living organisms are much older than human civilisations. At a time when 'saving' resource hungry, polluting industrial economies is presented by the elite as 'saving civilization', *Biocivilisations* shows us how we can learn from bacteria and fungi to create ecological civilisations, in harmony with all beings on Earth. Slijepčević invites us to shed anthropic arrogance and adopt ecological humility. 'More than 99.99% of the time that life has existed on Earth has been without us.' Microbes and plants have been around much longer that we have. They are our elders in the art of living. Their biocivilisations can teach us how to make the transition and paradigm shift we *must* make, for our future as a species and the future of life on Earth. This humbling transformation is the path of hope, to 'save ourselves, and the living planet, from our own violence.'

We need a new paradigm, a new science of life, for a respectful relation with nature. We need to discover our place and our work in the biosphere. We need a new humility to overcome anthropic arrogance. Belonging to the web of life, learning from other members of our Earth family on how we can live together in symbiosis, creates new possibilities, and new hope.

Predrag Slijepčević reveals the intelligence, creativity, autopoiesis and self-organisation of all living organisms. 'When our human universe opens to the universes of our planetary relatives,... we become citizens of a biological multiverse, who now embrace the future with a sense of wonder and responsibility never experienced before.' *Biocivilisations* is a guide for us to create a future in harmony with life on Earth.

—Dr Vandana Shiva

Preface

This book is about applying the science of life to discover civilisations developed by 'others'. Do animals practise science and art? Do trees exercise engineering skills? How can we listen to bacterial conversations and translate their 'thoughts' into human words? Does nature, indeed, possess a mind? These are some of the questions I have asked in search of non-human civilisations on our planet. I hope I shall not be seen as a practitioner of pseudoscience but as a scientist who tried to think differently and got some things right. Errors, when walking along unusual intellectual paths, are inevitable.

Because of the peculiarity of this task and its evolution as a personal project, the reader should know who I am and how my own experiences shaped the book. This may soften the scientific narrative, which is admittedly idiosyncratic at times because of difficulties associated with interpreting the non-human worlds that surround us and even determine us; ours is only one world in a multitude of interconnected biological worlds.

I was educated as a veterinarian at the University of Sarajevo, in what was then Yugoslavia, from 1982 to 1987. In those days, Yugoslavia was an open and cosmopolitan country. Its elites embraced Western culture but were not afraid to criticise it. At the same time, Yugoslav elites resisted Soviet influences but admired the positive values of Russian culture. Other influences on Yugoslav culture – especially the culture of its central republic, Bosnia and Herzegovina, and its capital, Sarajevo – included the Ottoman and Habsburg Empires, whose values mixed easily with modern influences. Sarajevo in the 1980s was a unique cultural hodgepodge, like a fairy tale from *Arabian Nights* combined with a local rock 'n' roll scene, countercultural comedy and independent film.

In such a unique cultural habitat, students were encouraged to ask questions that their Western colleagues would not dare to ask. For example, if we veterinarians truly want to help animals, why don't we study their psychology? All veterinary trainees know that animals have feelings and can show signs of mental distress. Our teachers did not have answers. Several years later, when I discovered the writings of Carl Gustav Jung, it occurred to me that Jung's concept of the collective unconscious, which links the human psyche with the animal psyche, should be an obligatory area of study in veterinary schools. In fact, official acknowledgement of animal consciousness would come decades later with the 2012 Cambridge Declaration on Consciousness.

After graduating in 1987, I started a Ph.D. in radiation biology and genetics. I wanted to know how radiation damages chromosomes, the cellular structures that carry genes. Intellectual curiosity stimulated by the unique Sarajevo and Yugoslav culture prompted me to study philosophy at the same time. I thought philosophy might help me understand things that science ignores.

Over the next three years, I combined experiments on chromosomes with reading the vast literature on genetics and philosophy. I defended my Ph.D. thesis in 1991, several months before Yugoslavia disintegrated in a tragic war. Thinking ahead, I had secured two possible exit routes to the West: a Fulbright scholarship for postdoctoral studies at the University of California, San Francisco, and a combined scholarship for the Universities of St Andrews in Scotland and Leiden in Holland. Because I already had a connection with St Andrews, I opted to stay in Europe. As American sponsors searched for me in war-torn Sarajevo through diplomatic channels, I assured them I was alive and well in Europe.

During my Sarajevo days, my scientific and philosophical projects developed to the point that I had clear ideas for the future. I intuitively understood chromosomes as unique structures that require special attention. They contain genes, but they are more than genes. My research, which took me from Sarajevo to the Universities of St Andrews, Cambridge, Leiden and Brunel over the next thirty years, focused on how chromosome structure and function might hold the key to an understanding of ageing and carcinogenesis.

The war prevented me from completing my studies of philosophy at the University of Sarajevo, but I've always continued to read philosophy. This gradually fused with deep reading of Yugoslav poetry, in particular the world-renowned Vasko Popa. Admired by the likes of Ted Hughes, Octavio Paz and Charles Simic, Popa created an epic philosophical vision that resembled 'a universe passing through a universe'. His universe was a naturalist one, enveloping plants, animals, people and things in a multitude of cosmic games, with uncertain but exciting outcomes.

Looking back from a thirty-year vantage point, it seems to me that the fusion of personal experiences, from my exposure to Yugoslav culture and my veterinary studies to the science and philosophy projects infused with poetry, combined with my research on chromosomes, made this book almost inevitable. I now offer it to you. I hope you will see, as I did, that when our human universe opens to the universes of our planetary relatives, a kaleidoscope of astonishing beauty emerges, concentrating four billion years of evolutionary history into a dizzying new vision. We become citizens of the biological multiverse, who now embrace the future with a sense of wonder and responsibility never experienced before.

INTRODUCTION

The Mystery of Life

What is life? This is the most important question in biology. And not only in biology. Physicists, chemists, mathematicians, anthropologists, philosophers and artists ask the same question and search for answers. Indeed, a physicist was the first to ask the question in a meaningful way and offer a route to an equally meaningful answer. Erwin Schrödinger's book *What Is Life?*, published in 1944, achieved cult status in the world of academia and beyond. He is credited with an originality of thought that resonates with scientists and fits particularly well with neo-Darwinian biology – a school of thought popularised by Richard Dawkins that merges Charles Darwin's and Alfred Russel Wallace's ideas of natural selection with genetic determinism.

Yet there is a growing body of scientists, philosophers and artists who do not share Schrödinger's vision of the science of life. The main disagreement lies in a misplaced reductionism. That is, if you believe in Schrödinger's ideas, life can ultimately be reduced to genes and information codes. This form of reductionism has become a ruling metaphor of our age. A football coach such as José Mourinho can say, with deep conviction, that winning is in his DNA. The superstar David Beckham can say that football is in England's DNA. And they wouldn't be wrong. The scientific vision of evolution, encapsulated in the famous tree of life, which has only three branches, is based on the DNA metaphor.[1]

However, the wisdom of our age is flimsy. It is becoming increasingly clear that life cannot be reduced to genes and codes. The gene metaphor is not only too simplistic but also deeply flawed. If there is a way to reduce life to a single principle, that principle must acknowledge the creativity of life that turns genetic and informational determinism on its head. Creativity and determinism are opposing forces. One force searches for novelty, the other suppresses it. The belief that changes in genes, or gene mutations, are the ultimate source of biological novelty is shattered by epigenetics – DNA is not the only messenger of biological information.[2] A genius mathematician, Freeman Dyson, summed up our misplaced obsession with genetic determinism: 'The rule of the genes was like the government of the old Hapsburg Empire…*despotism tempered by sloppiness*.'[3]

If we downplay the importance of genes and codes, we arrive at a different principle of life, formulated by anthropologists Anne Buchanan and Kenneth Weiss: 'Life is an orderly collection of uncertainties.'[4] This is a groundbreaking statement. Life suddenly breaks free from our deterministic prison and turns into a river that sweeps us along in its current on an indeterministic – that is, uncertain – journey. We suddenly realise that this journey is the biggest mystery in the universe. This is because the universe is an open system.[5] The openness of the universe reduces the importance of determinism, including genetic and informational determinism.

With the sobering thought that our science of life may be based on flimsy principles, I turn to the topic of this book. The term 'biocivilisations' is an acknowledgement of the mystery of life and its deep uncertainty, as opposed to the quasi-certainty of the human position governed by the narrow time window of the Scientific Revolution. Humans, and our technoscience, are too young, evolutionarily speaking, to be able to claim any form of wisdom. More than 99.99% of the time that life has existed on Earth has been without us. Given that all forms of life except bacteria become extinct and are replaced by new forms of life, it's clear that life will continue in some form, post–*Homo sapiens*, long into the future.[6]

A simple solution to the discrepancy between our evolutionary youth and the maturity of life and its wisdom is to turn anthropic naïvety on its head. Let's allow bacteria, amoebas, plants, insects, birds, whales, elephants and countless other species – all evolutionarily much more experienced than us – to lead the way and show us how to rectify schoolboy ecological errors that reduce the chances of human civilisation surviving for even the next hundred years. The consequence of this turning is the emergence of millions of 'new' civilisations that preceded our own. Some of these civilisations, which I call biocivilisations, are hundreds or thousands of million years old. Take bacteria, for example. Bacteria have built cities and connected them with information highways.[7] This process brought the whole planet to life 3,000 million years ago. The name of the first biocivilisation, which exists to the present day, is the Bacteriocene. This primordial biocivilisation has given birth to all other biocivilisations, including the most recent one, the Anthropocene.

One of the youngest biocivilisations (humanity) must communicate with other biocivilisations in search of the wisdom we so plainly lack. You may think this is impossible. How can we talk to elephants and whales? Or bacteria and amoebas? Yet imagination is a powerful tool. In a 1917 satirical short story, 'A Report to an Academy', Franz Kafka imagined a world in which an ape called Red Peter adopted human behaviour in order to escape from the zoo.[8] Red Peter was so successful at assimilating with human civilisation that he was invited to address the esteemed members of a scientific academy. Our task is to do, in all seriousness, what Red Peter did, but inversely. In the world of biocivilisations, humanity must seek to understand and adopt the best practices of other biocivilisations to the depth and degree that we can convincingly address the academy of life and its principal authority: Gaia. My argument is that this humbling transformation – from a self-centred and naïve young species into a more mature, desegregated species that is aligned with its surroundings – is the only true prospect for us to save ourselves, and the living planet, from our own violence.

Part I

BEYOND HUMANS

CHAPTER ONE

How to Build a Biocivilisation

Mind is the essence of being alive.

GREGORY BATESON[1]

The public often idealise science as a superior way of knowing, fortressed in objectivity and matter-of-fact rationalism. Yet what makes possible this way of knowing are the conglomerates of thirty-seven trillion cells that make up the human body, along with ten times as many microbes that enable our bodies to exist within the invisible microbial cloud that straddles the biosphere.

This is one of the mysteries of the world that remains unexplained: how this scientific rationalism – that over the last 300 years has taken humans from a modest earthly presence to the edge of the cosmos – emerged from the four billion years of microbial evolutionary games that Ernst Mayr, one of the most influential biologists of the 20th century, called 'stupid' in a famous dialogue with Carl Sagan.[2] In other words, how can we reconcile our superior intellectual rationalism with the perceived stupidity of microbes, whose wildness has been tamed and transformed, in the manner of well-trained Shakespearean actors, into powerful brain cells that can scan the cosmos, write love letters, compose the New World Symphony or paint *Guernica*?

Here is another question, this time directed at those who like to bet. Who is really stupid: people or microbes? Mayr wagered that microbes are champions of stupidity. He used an argument popular amongst futurists: high cerebral intellect always carries with it an existential risk.[3] Human technologies, from the steam engine to artificial intelligence (AI), expand our comfort zones enormously, but at the same time they endanger long-term survival. Mayr argued that microbes have neither technology nor intelligence. Their 'stupidity' has given them evolutionary longevity – four billion years of carefree monotony.

On the other hand, the microbiologist James Shapiro wagered that we humans are the champions of stupidity. In a paper entitled 'Bacteria are small but not stupid', he wrote: 'Bacteria are far more sophisticated than human beings in controlling complex operations.'[4] Shapiro's colleague, Eshel Ben-Jacob, further developed the idea of a collective bacterial mind.[5] The planetwide bacterial communication network – the bacteriosphere – makes life on Earth possible by controlling biogeochemical cycles of organic elements, even despite multiple evolutionary catastrophes and collapses.[6] According to Ben-Jacob, bacterial communication exhibits characteristics known to language experts, including syntax (language structure) and semantics (meaning).[7] The idea of the universal grammar that we associate exclusively with human language is, in fact, a replica of a much older school of communication.

Organism plus Environment

At first glance, Mayr's position is a progressive one. Science and philosophy teach us that humans are the most intelligent organisms in the history of life. Who would even bother to argue against an attitude so culturally ingrained? Shapiro's position, on the other hand, is counterintuitive. Evolution appears progressive: the complexity of organic forms increases with evolutionary time. It seems like microbes *must* be primitive relative to the complex biological designs of mammals, for example.

This book is an ante to Shapiro's bet. I will tell you a story of biological civilisations, or 'biocivilisations' for short, much older than human civilisation. The story of biocivilisations will show that bacteria, other microbes and all other forms of life are not stupid. On the contrary, the planetary bacterial superorganism, or the bacteriosphere, has been running the biogeochemical affairs on Earth for billions of years with a kind of intelligence that may forever remain beyond human capacities.

When Richard Feynman stated, 'What I cannot create, I do not understand', he captured the purpose of science in seven words.[8] The story of biocivilisations, based on the most recent scientific evidence, will create a testable model of life's true diversity, which is hidden from view by the anthropocentrism of mainstream science. This model may help us understand our place in a world turned upside down; a world in which microbes actually dominate, and animals, like humans, are temporary intruders, tricked onto a dangerous path by an unprecedented sense of self-aggrandisement. Gregory Bateson, a cyberneticist and philosopher, called this dangerous path 'pathologies of epistemology'. For him, even Darwin was wrong.

> *Now we begin to see some of the epistemological fallacies of Occidental civilization. In accordance with the general climate of thinking in mid-nineteenth-century England, Darwin proposed a theory of natural selection and evolution in which the unit of survival was either the family line or the species or subspecies or something of the sort. But today it is quite obvious that this is not the unit of survival in the real biological world. The unit of survival is organism plus environment. We are learning by bitter experience that the organism which destroys its environment destroys itself.*
>
> *If, now, we correct the Darwinian unit of survival to include the environment and the interaction between organism and environment, a very strange and surprising identity emerges:* the unit of evolutionary survival turns out to be identical with the unit of mind.[9]

Bateson precisely captured the meaning of the term 'civilisation': the relationship between organisms and their environments. The relationship is deeply knowledge-based, or mind-like. This natural epistemology is best represented by an analogy: imagine that one neuronal cell in the brain is an organism. The surrounding neuronal cells represent the environment. The 'wiring' between a single cell and its environment will lead to all the cells 'firing' together as part of brain activities.[10]

Biocivilisation, on the other hand, is the emergence of wiring between organisms and their environments that fills the biological world with meaning. Indeed, organisms are constantly challenged by their environments. Bacteria must find a new source of food, ants must protect their fungal gardens from microbial intruders, honeybees must invent more efficient algorithms for nectar collection, etc. Environmental challenges force organisms, or autonomous natural agents, onto the path of natural learning. Through the process of learning – the neuronal-like wiring of autonomous natural agents with their surroundings – organisms construct environments.[11] Ultimately, organisms and environments become indistinguishable from each other. For example, the human body is as much an organism (a corporate conglomerate of cells) as it is an environment (a platform for 37 trillion eukaryotic cells and 400 trillion microbes).[12]

In a biocivilised world, there is no difference between human beings and bacteria in the civilising and mind-like process of learning. The historian Niall Ferguson remarked: 'Civilizations are partly a practical response by human populations to their environments – the challenges of feeding, watering, sheltering and defending themselves.'[13]

One can also, however, replace the word 'human' with 'bacterial', 'animal', 'plant' or 'fungal' in the above quote. Nothing essential will change. In the world of biocivilisations, human civilisation is just a fragment of the huge spectrum of biocivilisations that together form the planetary biosphere. Applying the Batesonian analogy even further, the biosphere, or Gaia, is a composite biocivilisation capable of self-regulating, a unit mind. As Bateson remarked: 'Mind is a necessary, an inevitable function of the appropriate complexity, wherever that complexity occurs. But that complexity occurs in a great many other places besides the inside of my head and yours.'[14]

The Gaian mind, I will argue in the book, becomes apparent when we discover how individual biocivilisations that are 'wired' together, also 'fire' together. This enables Gaia, the planetary ecosystem, to regulate itself and survive the challenges brought about by periodic evolutionary catastrophes. Gaia is a fast-thinking and intelligent system, a phoenix that rises from its own ashes. In this system, biocivilisations come and go, but there is one constant. Bacteria have been around since the outset of life on Earth. Lynn Margulis argued that bacteria are the basic unit of Gaia.[15] Indeed, this basic unit may prove indestructible.

The Story

This book has three parts. Part 1, Beyond Humans, is about humanity's distorted views of the biological world and our so-called planetary dominance. This distortion is embodied in the term 'Anthropocene', coined by Nobel Prize–winning chemist Paul Crutzen and his collaborator Eugene Stormer to describe how 'humans have replaced nature as the dominant environmental force on Earth'.[16]

But, in fact, microbes are the dominant environmental force on Earth. Founders of the biosphere, their presence is virtually everywhere – including deep in the oceans, high on mountain peaks, amongst tropospheric clouds, in every kind of forest and even in our bodies – and it is a constant reminder that humans are non-essential by-products of what mostly amounts to microbial evolutionary games. Take humans out of the equation and nothing of significance would happen to the biosphere's natural trajectory. Indeed, many species would flourish in our absence. Others would go extinct. But none of this would cause the steady beating pulse of the biosphere to so much as skip. By contrast, take microbes out of the biosphere and the whole biosphere would collapse.[17]

The human thirst for self-aggrandisement – encapsulated in Yuval Noah Harari's new label for our species, *Homo Deus* – is a monumental self-delusion.[18] The biosphere existed without us for more than 99.99% of its history. We are far more likely to remain an evolutionary statistical error than God-like creatures. We have been deceived into delusional fairy tales from bestselling books that hawk the idea of human biological superiority, and we have been sold fantasies about human immortality that will enable us to control the solar system, or even the galaxy, in a not-so-distant future.[19] My argument, on the other hand, is that a humbling, not an aggrandisement, will yield our most valuable lessons, from microbes, fungi, plants and other animals – organisms with far greater evolutionary experience than us.

The 20th-century British mathematician and philosopher Bertrand Russell recognised delusions of this type decades ago; he called it 'madness' and was unafraid to point the finger at scientific hubris as the source of such insanity. 'The philosophies that have been inspired by the scientific technique are power philosophies, and tend to regard everything non-human as mere raw material. Ends are no longer considered; only the skilfulness of the process is valued. This also is a form of madness. It is, in our day, the most dangerous form, and the one against which a sane philosophy should provide an antidote', Russell asserted in his celebrated 1945 book *A History of Western Philosophy*.[20]

Viewing *Homo sapiens* as the dominant environmental force on Earth is not only a form of madness, it could not be further from the truth. The real force of life rests with microbes; the Bacteriocene runs the biological world. Microbial civilisation is the most powerful civilisation that has ever existed on this planet. In the remainder of this chapter, I will describe the basic elements of microbial biocivilisation – language, mind and memory – and show how almost all elements of human civilisation have precursors in the bacterial world.

In Chapter 2 (Against Mechanism), I will outline four principles of life that challenge mainstream biology – the science of life as subservient to physics. You will see that biology subservient to physics is a heavily biased science resulting from a wrong epistemology. In Chapter 3 (Pride and Prejudice), I will challenge some of our anthropic prejudices, from the way we interpret intelligence, to the fallacy of the Anthropocene and our understanding of sentience. I will argue that the attitude of modern biology is wrong. What we call life is a process that operates in a mind-like manner. Exactly as Bateson described: 'Mind is the essence of being alive.' We have to liberate biology from the grip of the mechanism and acknowledge that no organism is a machine. Each organism is a unit of mind – a holon integrated into the global Gaian mind. By contrast, the machine is mindless – a piece of dead matter that will never become alive, despite enormous efforts by futurists of the physicalist persuasion to convince us it will. The three chapters of Part 1 will underpin the central part of the book, Part 2 (Brave New World), the story of biocivilisations.

Part 2 is not an homage to Aldous Huxley and his futuristic novel, but rather a challenge to the kind of futurism that is based on mechanistic science. We have to be brave enough to recognise that many technological achievements, usually taken to be human inventions, existed in the world of biocivilisations long before the human species emerged in the span of evolution. Those biocivilisations can be witnessed every day in forests, fields, oceans and on mountains. The lives and civilisations of plants, termites, bacteria, birds, elephants, honeybees, ants, whales and octopi are replete with amazing and valuable technologies (Chapter 4, Civilising Force). Communicators (Chapter 5), engineers (Chapter 6), scientists (Chapter 7), doctors (Chapter 8), artists (Chapter 9) and farmers (Chapter 10) existed long before human civilisations emerged.

Finally, in Part 3 (Looking Forward), I argue for a new school of thought in the science of life, liberated from the stranglehold of a mechanistic world view. From Descartes onwards, all non-human organisms have been considered mere machines or, in the vocabulary of neo-Cartesians such as Richard Dawkins, dumb 'lumbering robots' controlled by selfish genes.[21] Indeed, most biologists are accustomed to using the terms 'mechanism' and 'mechanistic' to describe the inner workings or the behaviour of bacteria, archaea, protists, fungi, plants and non-human animals. Although there is some value in reducing biological processes to mechanistic explanations, and there is a promising possibility of identifying common ground between biocivilisations and the mechanistic view of life, this reconciliation must grow from a recognition that life is more than mechanism.

Microbial Language

Ever since life emerged on Earth, there has been constant communication amongst organisms. But this communication is deceptive. These are no talking heads, but the communication is not unlike our language. Instead, organisms communicate through exchanging biological signals and semiotic signs – chemical messages, electrical impulses, scents, body movements, etc.

Bacteria were the first organisms to speak up. Unfortunately, we are deaf to their conversations. Bacteria 'talk' to each other and to all other organisms on the planet. For example, bacteria from our microbiome – the collection of bacteria, viruses, archaea and fungi inside and on the surface of our bodies – talk to us behind our backs. They bypass the conscious part of our intellect and talk to the unconscious. Scientists have discovered 'conversations' between bacteria living in our guts and cells in our brains. This type of talk is called the gut–brain axis. Bacteria from our guts direct our brain cells to secrete serotonin, which improves mood.[22] Gut bacteria drug our nervous systems without us even being aware that it's happening.

Language is a set of symbols that convey meaning. Every linguistic sign reflects the superiority of mind over matter. When we utter a word, we launch a non-material abstraction full of meaning into the semiotic stratosphere. There, our symbols mix with bacterial, viral, plant and animal symbols in true Tower of Babel fashion, with the crucial difference that the number of languages in the semiotic stratosphere is far greater than in the biblical story. This fascinating biosemiotic construction intrigued the celebrated writer Umberto Eco, who was once impressed by the thought of a biosemiotician friend: 'Instead of thinking whether cells speak like us, the question should be asked whether we speak like cells'.[23]

How do bacteria talk to each other? James Shapiro discovered their semiotic symbols – i.e. the 'words' of bacterial language. He identified a large group of chemicals that bacteria exchange in communication with each other.[24] There are linguistic chemicals used in the communication of bacteria of the same species. There are also linguistic chemicals exchanged between bacteria of different species, in the manner of constructed international languages such as Esperanto or the true global language, English.

Conversations between bacteria and people, or bacteria and plants, are conducted in a language that scientists call cross-kingdom communication. The most successful communicators, including cross-kingdom communicators, are bacteria – true biological polyglots. Bacteria speak and understand all the languages of the world, from their own mother tongues, to plant and animal languages, to cross-kingdom communication that reverberates with the biosphere in the most complex music the universe has ever known. Bacteria also talk to viruses, semi-living biogenic structures, by detecting the words of a viral language – only recently discovered – based on arbitrium, which consists of a peptide composed of six amino acids.[25] Thus, mostly thanks to bacteria, the biosphere becomes the semiosphere – a compendium of biological signs and the domination of mind over matter.

The Microbial Mind

We know from experience that words don't make sense without the mind to decode and interpret them. The mind unites spoken words into sentences, then into stories, or perhaps into algorithms (if words are replaced by mathematical symbols). Without the mind, there is no storytelling, and this is also true of the biosphere and the Gaian mind. In other words, if bacteria have no mind, their talk is pointless; it represents little more than a form of mindless chatter. Mainstream biology, in a true Cartesian manner, does not allow for the existence of a bacterial mind. Even when leading scientists are willing to get into details of bacterial language, the idea of a bacterial mind controlling bacterial language is considered off-limits.[26]

Non-conventional scientists such as Eshel Ben-Jacob, Gregory Bateson and Lynn Margulis, however, have argued against the shortsightedness of mainstream biology, and especially against the prejudices that scientists cultivate towards non-human organisms – a kind of evolution-based anthropic racism rooted in modern culture. Ben-Jacob developed the concept of the 'bacterial brain', which is complete only when combined with Bateson's concept of the natural mind.

Ben-Jacob argued, similarly to Shapiro, that bacteria are multicellular communities (or colonies), with a typical colony consisting of 10^9–10^{12} individual organisms, and with the entire bacterial population connected into a global bacterial superorganism or bacteriosphere. Bacterial colonies are constantly using language to solve the problems presented by their environments. A colony will assess a problem – for example, food shortages – through the collective examination of the environment and collective gathering of information using bacterial language. Once the nature of the problem is determined, bacterial colonies use information about past problems, stored in the colony's collective memory. In this manner, the colony begins distributed information processing to solve the emerging problem. The problem-solving process transforms the colony into a structure most similar to the human brain. Bacterial colonies become 'super-brains' that perform acts of natural computation.[27] When we look from this perspective at the planetary bacteriosphere, in which bacterial colonies are connected through bacterial language – a living equivalent of the internet (see 'The Internet of Living Things', page 16) – we see glimpses of a planetwide bacterial brain that has maintained biogeochemical balance for billions of years.

Gregory Bateson would probably agree that a bacterial colony constitutes a form of the natural mind. Bateson often reminded his audience that the mind exists in nature in many more places than just inside our heads. To assess whether a biogenic structure meets the requirements of the natural mind, Bateson applied six criteria[28]:

1. The mind is the unity of the parts that communicate with each other;
2. The mental process creates feedback;
3. Communication between parts is driven by the 'difference that makes a difference' or biological information;
4. The mental process requires energy;
5. Biological information directs changes in the physical environment;
6. Biological hierarchy is constrained by both bottom-up (cells to ecosystem) and top-down (ecosystem to cells) forces.

Bacterial colonies and the planetary bacteriosphere meet all of Bateson's criteria of the natural mind. More than a century prior, Darwin stated the following about the mind: 'The difference in mind between man and the

The Internet of Living Things

Creating a huge global network connecting billions of individuals might be one of humanity's greatest achievements, but microbes beat us to it by more than three billion years. These tiny single-cell organisms aren't just responsible for all life on Earth, they also have their own versions of the World Wide Web and the Internet of Things. Here's how it works.

Much like our own cells, microbes treat pieces of DNA as coded messages. These messages contain information for assembling proteins into molecular machines that can solve specific problems, such as repairing the cell. But microbes don't just get these messages from their own DNA. They also swallow pieces of DNA from their dead relatives or exchange them with living mates. These DNA pieces are then incorporated into their genomes, which are like computers overseeing the work of the entire protein machinery. In this way, the tiny microbe is a flexible learning machine that intelligently searches for resources in its environment. If one protein machine doesn't work, the microbe tries another one. Problems are solved through trial and error.

But microbes are too small to act on their own. Instead, they form societies. Microbes have been living as giant colonies, containing trillions of members, since the dawn of life on Earth. These colonies have even left behind mineral structures known as stromatolites. These are microbial metropolises, frozen in time like Pompeii, that provide evidence of life from billions of years ago. Microbial colonies are constantly learning and adapting. They emerged in the oceans and gradually conquered the land – and at the heart of their exploration strategy was information exchange. As we've seen, individual members communicate by exchanging chemical messages in a highly coordinated fashion. In this way, microbial society effectively constructs a collective 'mind'. This mind directs pieces of software, written in DNA code, back and forth between trillions of microbes with a single aim: to fully explore the local environment for resources. When resources are exhausted in one place, microbial expedition forces advance to find new lands of plenty. They transmit their discoveries back to base using different kinds of chemical signals, directing microbial society to transform from settlers to colonisers. In this way, microbes eventually conquered the planet, creating a global microbial network that resembles the World Wide Web, but using biochemical signals instead of electronic-digital ones. In theory, a signal emitted by bacteria in waters around the South Pole could travel almost instantaneously to bacteria in the waters around the North Pole.

The similarities to human technology don't stop there. Scientists and engineers are now working on expanding our own information network into the Internet of Things, integrating all manner of devices by equipping them with microchips to sense and communicate. Your fridge will be able to alert you when it is out of milk. Your house will be able to tell you when it is being burgled. But microbes built their version of the Internet of Things a long time ago. We can call it the Internet of Living Things, better known as the biosphere. Every organism on the planet is linked in this complex network that depends on microbes for its survival. More than a billion years ago, one microbe found its way inside another microbe that became its host. These two microbes became a symbiotic hybrid known as the eukaryotic cell, the basis for most of the life forms we are familiar with today. All plants and animals are descended from this microbial merger and so they contain the biological 'plug-in' software that connects them to the Internet of Living Things.

For example, humans are designed in a way that means we cannot function without the trillions of microbes inside our bodies (our microbiome) that help us do things like digest food and develop immunity to germs. We are so overwhelmed by microbes that we imprint personal microbial signatures on every surface we touch.

The Internet of Living Things is a neat and beautifully functioning system. Plants and animals live on the ecological waste created by microbes, while to microbes all plants and animals are, as author Howard Bloom puts it, 'mere cattle on whose flesh they dine', whose bodies will be digested and recycled one day.

Microbes are even potential cosmic tourists. If humans travel into deep space, our microbes will travel with us. The Internet of Living Things may have a long cosmic reach. The paradox is that we still perceive microbes as inferior organisms. The reality is that microbes are the invisible and intelligent rulers of the biosphere. Their global biomass exceeds our own. They are the original inventors of the information-based society. Our internet is merely a by-product of the microbial information game initiated three billion years ago.[29]

higher animals, great as it is, certainly is one of degree and not of kind.'[30] Bateson, however, was much more radical. For Bateson: 'Mind is the essence of being alive.' Interestingly, Bateson considered Lamarck to be the greatest biologist in history, not Darwin. Presumably because, amongst other things, Lamarck sensed the intelligence of bacteria.

Microbial Memory

The collective memory of microbes goes back almost four billion years to the moment when these tiny and invisible organisms emerged on Earth. How is this possible? The first line of memory is the microbial genome that stores blueprints for the oldest protein constructs. This is no exaggeration but simply a consequence of the biological postulate of vertical gene transfer. Since bacteria and archaea were the first organisms in the history of life, all organisms that evolved from them, including humans, share their four-billion-year genetic heritage.[31]

But genes are only one line of memory. The other line of memory is much more important: the organism as a biological construct that incorporates genes. Genes are important, but they are secondary. The emblem of neo-Darwinism – 'the selfish gene' – is becoming an obsolete concept. How can a gene be selfish when it lacks a self? This is a question Lynn Margulis asked Richard Dawkins at a meeting. No one has yet come up with a convincing answer.

Margulis argued that the basic unit of life, and therefore of memory, can only be the simplest cell – a microbe such as a bacterium or an archaeon. Each bacterium is an open thermodynamic system that exchanges matter, energy and information with its environment. Each bacterium has an instinctive sense of its own body in the context of the external environment. It has perception, memory and the ability to make decisions, plan the future and communicate. These characteristics are to be expected from a biological system that has a sense of its own body.

A gene or a piece of a DNA molecule, on the other hand, is a simple biological code that serves the bacterium as an aid in the transmission of biological information, and in the process exchanges matter and energy with the environment. This code is meaningless without the context of the bacterial 'body' as an open thermodynamic system. The DNA code is a form of bacterial 'thought' – a hypothesis subjected to an evolutionary test.

Interestingly, the source of genes, which are biological 'thought' material, is not only bacterial or archaeal genomes, but viruses, plasmids, naked genes and other DNA pieces involved in horizontal gene transfer (HGT). Some scientists argue that viruses may be a precursor to life – a position supported by the fact that bacteria, the most dominant form of life in the biosphere, cannot exist without viruses.[32] The bacteriosphere needs an auxiliary biogenic structure. Scientists call this structure the virosphere. The virosphere and the bacteriosphere are the foundations of life. The most numerous viruses in the virosphere are those that infect bacteria (although bacteria developed a system of defence against unwanted viruses, or, metaphorically, unwanted 'thoughts', so that they can preserve their own 'common sense').[33]

Eshel Ben-Jacob defines the nature of bacterial memory as follows: a combination of (a) internally stored information in the genome of each bacterium and (b) information that the bacterial society, which makes up the colony, collects from its environment and stores in the structure of the colony. In other words, genetic memory by itself is not enough for the process of wiring bacteria to the environment. Genetically stored information only serves to trigger more complex collective information-processing abilities, which then create new knowledge that bacteria need to learn about new conditions in the environment. Thanks to this memory, the bacterial colony turns into a brain-like entity capable of natural learning, and thus becomes a form of Bateson's natural mind. The conclusion is self-evident: the ability of bacterial colonies to remember past events transforms the planetary bacteriosphere into the collective memory of the biosphere.

No matter how much *Homo sapiens* might deny the dominance of microbes on Earth, and artificially impose human dominance, reality refutes us. The COVID-19 pandemic is one example amongst many that ruthlessly revealed the holes in our understanding of biological reality. The bacteriosphere and virosphere are the basis of life (see Chapter 2). Although we can't deny the usefulness of the term in revealing the destructiveness of human impact, the Anthropocene is little more than an anti-Copernican delusion of modern civilisation.

The First Scientists

But how did humans get to a point where we are so enamoured of a faulty interpretation of the biological world – the delusion of the Anthropocene? It may sound paradoxical, but the delusion was inspired in part by some of the greatest achievements of modern civilisation, many of which occurred during (or would not have occurred without) the Enlightenment. Enlightenment thinkers introduced a sweet but dangerous idea: *Homo sapiens* is the most sophisticated organism in the history of life. Human sophistication has no limits. Transhumanism and man-made machines will help us conquer the cosmos and transform us into a cosmic deity.[34]

Today, the term used to justify the superiority of the human mind is 'post-biological evolution' – the idea that humanity and its machines will dominate the world. Or perhaps that machines will turn against humanity and establish mechanical dominance of the world through the Machinocene (a term coined by Huw Price on the portal *Aeon*) or superintelligence.[35] I suspect Bateson would have questioned this possibility. He thought of machines as mindless. Indeed, in my opinion, post-biological evolution falls into the same category of delusion as the Anthropocene.

Of course, the Enlightenment was enormously useful for the development of civilisation. Without the Enlightenment, there would be no science in the modern form. However, one should also acknowledge, in the manner of far-sighted thinkers such as Bateson, Margulis and others, its major fallacy: anthropic arrogance towards all other life forms, especially microbes. At its worst, this arrogance becomes a form of anthropic racism – our technologies are responsible for a biocide of planetary proportions, recently dubbed 'biological annihilation' or even the 'Necrocene'.[36] The fallacy, if one can rely on psychiatry, may be a consequence of the unresolved Copernican complex. Although Nicolaus Copernicus, the 16th-century Polish priest turned scientist, showed us that we live on an insignificant planet that orbits around an insignificant star, we have not given up the comforting but dubious idea that *Homo sapiens* is the peak of evolution around which the entire cosmos revolves. This is apparent, for example, in the concept of the anthropic argument that places humanity at the centre of the thinking universe.[37] Anthropic exaggerations, despite the rationalism of the Enlightenment and Copernican caution, only serve to add weight to the validity of Shapiro's position. Actually, Shapiro's position is probably the best-articulated form of Copernicanism in biology.

If we adopt Copernicanism in biology – the position that there is nothing special about the human species – then the only relevant yardstick of the two competing options – Mayr's deeply anthropocentric option and Shapiro's biocentric one – is evolutionary scale. Although we are unable to experience the temporal scale of evolution, we have ample evidence to conclude that *Homo sapiens* is a young and inexperienced species whose modern civilisation is not yet aware that, in Bateson's words, by destroying its environment, it destroys itself. The same thesis in a different context was advocated by Chief Seattle: 'Man did not weave the web of life, he is merely a strand in it. Whatever he does to the web, he does to himself.'[38] Likewise, Lynn Margulis ended her book *Symbiotic Planet* with a passage containing the following sentence: 'We cannot put an end to nature; we can only pose a threat to ourselves.'[39]

Although Shapiro's position is not anthropocentric, it is deeply humanist in a nuanced and particular sense – it accepts humanity's place in the world. This form of humanism allows for a constructive criticism of Cartesian science, as well as of the Enlightenment itself. As I will show in this book, microbes have been practising science (as well as engineering, farming, art, medicine and more) since long before the Enlightenment; indeed, long before the appearance of *Homo sapiens* at all.

The Artful Science

Flipping the prevailing narrative has two purposes that will make the journey into the world of biocivilisations more comfortable. First, as Gregory Bateson thought, turning the human world upside down is a valuable metaphor that can help us understand our place in the world better. The consequence of this metaphoric turning is not trivial. It may help us realise that we are lost in the jungle of a wrong epistemology, without being aware of how this causes us to misconstrue the human place in the world.[40]

Second, it can help us find a road into the evolutionary past on which we can discover the origin of biocivilisations, and by doing so identify a way out of the jungle of wrong epistemology. The key person here is Lynn Margulis and her serial endosymbiosis theory (SET).[41] SET interprets life as an interconnected mosaic of living organisms that together form the Gaian superbiosystem, one in which viruses coevolve with bacteria, bacteria and archaea complexify into protists, protists complexify into plants and animals, and all these organisms integrate into the Gaian mind (see Chapter 2). SET allows us to argue that the mind-like structural coupling between organisms and environments that sparked the emergence of biocivilisations started with, and is still dominated by, microbes.

Kenneth Clark, a British art historian and broadcaster, remarked at the beginning of his hugely successful TV series *Civilisation*: 'What is civilisation? I don't know. I can't define it in abstract terms yet. But I think I can recognise it when I see it.'[42] Clark was referring to art as the peak of human civilisation. Lynn Margulis was able to discover for us the art behind biocivilisations, which in many respects surpasses everything human civilisation has ever produced, and likely everything it ever will.

Likewise, I view Gaia as an artistic system, one that constantly creates itself in the autopoietic sense described by Humberto Maturana and Francisco Varela: by encouraging viruses here and there, nudging plants from one habitat into another, attracting bacteria and fungi to an insect ecosystem, or directing big changes in its body by converting evolutionary catastrophes into new forms of biogenic art.[43] Gaia is the totality of all biocivilisations that defies the mechanistic view of life increasingly imposed on us by AI enthusiasts and their physicalist view of the world. We can call the totality of the Gaian mind the cognitive multiverse – a truly vast artistic system in which mind is more than computation.[44] Just as the sculpture of David is the product of an artist's mind, the oxygen-rich atmosphere is the product of an artistic spark that occurred when cyanobacteria discovered how to eat the Sun. Ever since then, Earth's atmosphere has been a huge planetary canvas on which artistic sparks thrive, leading to an endless cascade of new plant and animal biocivilisations that constantly enrich the complexity of the Gaian mind.

When we merge Bateson's and Margulis' ideas, we get a huge evolutionary theatre of perpetual mind-like mergers of organisms and environments over the course of four billion years. The name of this evolutionary play is biocivilisation. It consists of countless streams – individual biocivilisations that represent the cognitive spaces of individual species – joined together into the big river of the Gaian mind.[45] We humans are not able fully to capture or copy the art behind the Gaian mind. We can only admire it by trying to discover, like Bateson, Margulis, Shapiro, Ben-Jacob, Maturana, Varela and others, the streams in the river that can take us out of the jungle of mistaken epistemology.

Allow me to recapitulate: if you want to understand a biocivilisation, you must try to build a model of it in your mind, as Feynman so aptly described: 'What I cannot create, I do not understand.' My attempt to build a model of biocivilisation has revealed its three crucial elements: language, mind and memory – essential elements in all biocivilisations that have appeared as streams in the Gaian river ever since microbes started to merge.[46]

But building a biocivilisation with a human model offers only a limited glimpse into the complexity of the Gaian mind. Nevertheless, this is perhaps what true humanism represents. This is the humanism that makes life worth living – by catching glimpses of the Gaian mind, we integrate into it. We become part of the huge river that started flowing four billion years ago. Gaia is neither a mechanical system nor a giant biological machine. The next time you meet passionate adherents of modern computational concepts such as the Machinocene, AI-based superintelligence or post-biological evolution, please tell them that Gaia is a Heraclitean river that defies the mechanics of physicalism (see Chapter 2). Their machines will simply sink to the bottom of the Gaian riverbed. There, they will serve as meals for countless microbes – those that can eat even metals. After a while, these microbes will return the digested wreckage to the immortal Gaian flow.

For this reason, I have to correct myself. I said earlier that every single piece of evidence that I will use in the book will be based on science and this will make the book scientific. This remains the case. However, the way I will use the evidence to lead you through the story of biocivilisations will defy the classical science that is traditionally dominated by physicalism. I will use scientific evidence to integrate scientific evidence into the artistic Gaian mind. Inevitably, this process may offend hard-core mechanistic scientists. My angle, however, is different. It represents a badly needed avenue for the reconciliation of arts and sciences, whose relationships are poisoned by the high priests of both realms. A scientist-turned-poet friend articulated this false dichotomy in an interview on my blog: 'Science, art, music and literature explore the same world but using different tools.'[47]

My goal is to show that the seemingly separate worlds of art and science are, in reality, one, but perhaps with different faces, like the Roman god Janus. Both faces are essential. The scientific face will help us explore and survive our increasingly uncertain future, but only if we tone down the mechanistic reductionism. The artistic face will always remind us that we are not gods but tiny parts of the Gaian mind – a truly creative system that, like an artist, cannot predict its next iteration.[48]

So, the story of biocivilisations perhaps represents a form of artful science – the science that escapes the grips of physicalism and finds its true habitat: biology. This also means we cannot fully understand life through a physicalist science that mechanises everything from atoms to biosystems. This mechanophilia, or love for the machine, is a symptom of the 'pathologies of epistemology'. The only antidote to this pathology is the true science of life, a description of which you will hopefully find in the pages of this book.

CHAPTER TWO

Against Mechanism

Biology is fundamentally and qualitatively different from physical science.

WALTER ELSASSER[1]

The gist of Chapter 1 was to argue for a new biology liberated from the Cartesian prison, whose modern Praetorian Guards are those biologists who fancy a mechanistic view of life. The key manifestation of mechanism in biology is reductive materialism.[2] Most biologists interpret genes as pieces of biological software, cells as molecular factories, organisms as organic robots and nature as a giant mechanical system accountable to physics and mathematics. James Watson, a Nobel Prize–winning biologist, nicely summarised the mainstream view: 'There is only one science, physics; the rest is social work.'[3]

But there is strong opposition to the mainstream view. No mathematics and physics can ever fully explain biology, according to a growing body of non-mechanistic thinkers. Instead, biology will forever remain epistemologically open, not closed. The Gaian mind is the Heraclitean river that resists mechanism. Only a biology based on reductive materialism could declare microbes stupid. This is as much a manifestation of a wrong epistemology as it is a manifestation of our collective guilt caused by the unresolved Copernican complex. No matter how hard we try to justify the mistaken picture of the biological world, applying the flawed concept of the Anthropocene, the appalling reality of biological annihilation, the Necrocene and other manifestations of the pathologies of epistemology, it only serves to make bigger the elephant of the Copernican complex in the room.[4]

Many scientists will not like this argument, yet I am convinced that it will help us discover huge cracks in the edifice of mechanised biology. This edifice, dominated by the sense of mechanophilia, represents a greater threat to us than all bacteria and viruses in the world combined. In the beautiful economy of the Gaian mind, viruses and bacteria are here to help us, not to kill us. The microbial biocivilisation is the mother of all biocivilisations. Lynn Margulis put this biological truth beautifully: 'We can no more be cured of our viruses than we can be relieved of our brains' frontal lobes: we are our viruses.'[5] Modern genomic research gives credence to Margulis' thought: a large part of the human genome is made up of sequences that once belonged to viruses.[6]

The Great Escape

In this chapter, I will introduce key principles underpinning the concept of biocivilisations. You can probably guess that these principles will target mechanism and its inadequacy in the context of biology; this is the great escape from the Cartesian prison. Let me be clear from the outset: I am not blaming physics and physicists for this inadequacy. I am blaming biologists who, by insisting that biology should be subservient to physics, prevent the Copernican turn in biology that has already occurred in physics. The term 'mechanism', as I use it in this book, reflects a position deeply ingrained in academic biology – in particular molecular biology and genetics – that the living world is explicable in mechanistic, reductionist and deterministic terms.

In fact, physicists are far more supportive of a Copernican turn in biology than biologists are. For example, the physicist Walter Elsasser acknowledged: 'Biology is fundamentally and qualitatively different from physical science.' He was careful to avoid the label of vitalist. As a physicist, he knew that the laws of nature are equally applicable to organic and inorganic matter. Neither did he question the explanatory apparatuses of physics; he was merely stating that the apparatuses that work in physics may not work in biology. In other words, the structure and behaviour of organisms may not be reducible to physico-chemical causality.

An atom of hydrogen is the same on Earth as it is on Mars or Venus. However, there are no two organisms in the world that are exact copies of each other. Even identical twins are not the same individuals.[7] Even your own clone will be different from you.

We can see subtle differences between individuals even in the microbial world.[8] When a bacterium divides and produces a copy of itself, the original and the copy are not identical. One bacterium will contain damaged and imperfect molecules and will produce a progeny of ageing bacteria. The other bacterium will contain a set of undamaged and functional molecules and will produce a progeny of bacteria that will revitalise the population. Many biologists still think of bacteria as biological machines that do not age, but Elsasser argued that biological systems are fundamentally different from machines.[9]

Another physicist who interpreted biology in non-mechanistic terms was Nicolas Rashevsky.[10] For him, the fundamental property of biological systems was their organisation, which is independent from the material particles that make them up. Cartesian and Newtonian explanations may not be applicable to biological systems. Rashevsky's Ph.D. student, the theoretical biologist Dr Robert Rosen, further developed this line of inquiry. His position is best summarised in his own words from his 1991 book, *Life Itself*: 'It may perhaps be true that the question "What is life?" is hard because we do not yet know enough. But it is at least equally possible that we simply do not properly understand what we already know.... The question "What is life?" is not often asked in biology, precisely because the machine metaphor already answers it: "Life is a machine".'[11]

The mechanistic attitude that dominates academic biology is understandable. Scientists, including biologists, are born physicalists. As Stuart Kauffman memorably put it, scientists remain 'children of Newton' – a term that reflects the prevailing dogma of scientific training and education whose motto could be described as 'qualitative is nothing but poor quantitative'.[12] In other words, reductive materialism remains the bread and butter of science; if you are not a reductive materialist, you are not a scientist.

Even quantum mechanics, which signalled the end of the universality of Newtonian physics, did not translate into an end of physicalism. When Newton's children began dismantling the objects of their interest – the material fabric of nature – they were eager to play with these new tiny, shiny toys, micro-machines made up of atomic and subatomic components. But the joy that typifies play eluded them. They experienced a shock from which physics has never fully recovered. Instead of tiny mechanical toys fully explicable by mathematics, Newton's children encountered the pulsating hearts of micro-beasts that stubbornly pass through the net of reductive materialism. As Richard Feynman famously declared: 'I think I can safely say that nobody understands quantum mechanics.'[13]

That is to say, mechanical explanations disintegrate in the realm of the smallest. At the micro level, mathematical finality and precision are replaced by a cloud of statistical fuzziness. There seems to be more to the qualitative than reductionist science can account for. Robert Rosen expressed this view in a provocative manner, one that still upsets many mechanistic scientists: 'Why could it not be that the "universals" of physics are only so on a small and special (if inordinately prominent) class of material systems, a class to which organisms are too general to belong? What if physics is the particular, and biology the general, instead of the other way around? If this is so, then nothing in contemporary science will remain the same.'[14]

The principles of non-mechanistic biology I want to introduce are not at all obscure or poorly known. As a matter of fact, they are well known and well articulated. There are scientific journals and textbooks based on them. Conferences are organised and books published to promote them. Yet these principles remain isolated from each other and removed from the mainstream discourse that shapes academic biology. The four principles of non-mechanistic biology include:

1. Universal flux (*panta rhei*)
2. Agency (purpose and desire)
3. Symbiosis (living together)
4. Mind (hyperthought)

These four principles overlap and cannot be easily disentangled from each other. One can also view them as different aspects of a single principle of life (biocivilisations). Let's examine each principle separately.

Universal Flux (*Panta Rhei*)

The key principle of a biology liberated from artificially imposed subservience to physics is that life is neither mechanism nor thing. Instead, life is permanent change. It is a persistent process that eludes the reductionism and determinism of physics. If we accept this principle as the starting point for understanding life, we are no longer constrained by the dogmas of modern science (qualitative is just poor quantitative or, more precisely, biology is reducible to physics).

A discipline that views life as permanent change is known as processual biology. Its main tenets are the doctrine of universal flux, dubbed *panta rhei* ('everything flows') by Heraclitus and the Greeks, the process philosophy of Alfred North Whitehead and the research programme of organicist biologists in the first half of the 20th century, which have recently been rearticulated in the 2018 book *Everything Flows*.[15]

According to processual biology, the living world represents a hierarchy of processes rather than a hierarchy of things. Although molecules, cells, organisms, populations and ecosystems may look prima facie like things, processual biology interprets them as stabilised processes that are actively maintained at appropriate timescales. Processes persist, and by persisting they enable other processes. As a result, all biological processes are interconnected. It is enough to enter one process – for example, the development of an animal body – to be able to flow through all the processes that existed in the history of life, if you use the correct scientific approach and if you have enough time to execute it.

This is a radical departure from traditional substance ontology, also known as materialism, which has dominated Western thought since Aristotle. The bias towards things, which Alfred North Whitehead called 'dogmatic common sense', shapes our world view.[16] We are addicted to the material interpretation of the world. Science as we know it, based on principles of materialism and reductionism, is a direct consequence of this.

In processual biology, however, matter – the concrete things that you can see, touch, eat or manipulate – although important, is secondary. Life is the flow of matter through processes, governed by constraints.[17] It resembles the flow of water through the Heraclitean river. Water, the material or substance, is less important than its flow. Once the flow of matter through the process is stabilised, it will persist, as long as it has something to facilitate the persistence. For example, the energy of the Sun has facilitated the persistence of universal flux since the dawn of life.

There are processes in the non-living world, like whirlpools, tornadoes and flames, and you can demonstrate some of these, such as the Belousov–Zhabotinsky reaction, in a laboratory. Apart from a few exceptions, such as clouds and weather systems, processes in the non-living world tend to disappear quickly. By contrast, life on Earth is a compendium of processes that has been persisting for the last 3.8 billion years. There has never been a single total interruption in the process of life (which would have meant its extinction). Of course, some processes cease persisting: organisms die and species go extinct. But the Heraclitean river of life has never stopped flowing. It may have entered narrow passages, in the form of mass extinctions, but these were often followed by new and wide openings, like the Cambrian explosion.

The key property of processes is that they extend in space and time, as opposed to static entities arrested in space and time. A stone cannot extend itself by the process of reproduction. A book cannot imagine the future. However, the concept of space–time in physics, which fuses the three spatial dimensions with time into a single, four-dimensional manifold, offers a different story. According to processual biology, the four-dimensional manifold is already very similar to process ontology. Similarly, quantum mechanics has shown us that the material fabric of the world dematerialises at the quantum level. A particle has an alter ego; it also behaves as a wave. Thus, quantum mechanics is also very close to processual biology (but for the fact that it really should not be called 'mechanics').

Biology is naturally inclined towards the process ontology. For example, all organisms must take in matter to maintain themselves. This is called metabolic turnover – organisms are open systems that constantly exchange matter, energy and information with their surroundings. In the language of processual biology, organisms are stabilised processes through which matter, energy and information flow in an organised manner. When you add to this the fact that organisms experience life cycles, the argument for the processual nature of life becomes even stronger. Life cycles are a testament to the processual motto *panta rhei*. From the moment it comes into existence, an organism will change progressively. This means that stabilised processes have their own dynamics. Everything starts with birth, continues with development and ends with thermodynamic equilibrium, or death. There are no immortal organisms or endless individual stabilised processes. (Not only do organisms die, but species go extinct. The only 'species' that have persisted since the dawn of life are bacteria and archaea, in the form of ever-changing societies made up of mortal individuals. But bacteria and archaea defy the species concept, as I will argue in Chapter 5.)

Finally, no organism is an island. Life is a compendium of ecological interdependencies. Take bacteria out of the world and we will soon be dead; we need the bacteria in our microbiome to metabolise food into nutrients, assist our immune system and fight pathogenic microbes, and plants need bacteria to produce oxygen. The rule of life is symbiosis, or the principle of living together (see my principle 3). In the language of processual biology, no stabilised process is isolated from the symbiotic web of processes that together form the biological world. Gaia can be viewed as the 'meta' compendium of processes. Individual processes end, but Gaia is endless. Its flow – called homeorhesis by Lynn Margulis – is immortal.[18]

The best way to demonstrate the processual interconnection is detailed in the sidebar 'Flora or Mona Lisa?', page 32. These famous portraits – *Mona Lisa* by Leonardo da Vinci and *Flora* by Giuseppe Arcimboldo – are symbolic of different visions of biology. Mona Lisa symbolises a biology dominated by the substance ontology. Our explanatory apparatus, shaped by powerful cultural habits, interprets La Gioconda, or Lisa Gherardini, as a biological object independent of her environment. In today's world, dominated by AI and the neo-Darwinian interpretation of organisms as 'lumbering robots', Mona Lisa might even be thought of as a biological machine. Flora, by contrast, is a symbol of processual biology. The parts that form her are stabilised processes, dependent on the other processes that comprise her body. And, actually, these processes extend from Flora's body into her surroundings. You could also say that Flora is emerging from other life processes.

When we include time in the metaphoric interpretation of these two portraits, the differences between the visions of biology that they represent become even greater. Mona Lisa appears to most people as a biological object arrested in time. There are no hints in the portrait that would indicate her biological origins, or her potential future. By contrast, Arcimboldo's composite and ever-changing Flora is telling us that she is an exhibit in the living museum that emerged 3.8 billion years ago. Flora not only extends deep into evolutionary time, but also into the future.

Our sensory apparatus is blind to the processual nature of life. Brains serve to rapidly assess the ever-changing interactions between animals and their surroundings.[19] The analytical nature of our intellect, combined with some cultural prejudices, depicts the nature of reality in a heavily biased manner. We tend to focus on things and largely ignore processes. Physicists were the first to spot the weaknesses of our intellectual habits through the theory of relativity (e.g. the concept of space–time as a four-dimensional manifold) and quantum mechanics (dematerialisation at the level of the smallest). Niels Bohr was correct when he said: 'Those who are not shocked when they first come across quantum theory cannot possibly have understood it.'[20] Processual biology is telling us, loudly and clearly, that biology is ripe for the shock that physics experienced a century ago.

Flora or Mona Lisa?

The boundaries between the parts and the whole are almost non-existent on the canvases of the Renaissance painter Giuseppe Arcimboldo (1527–1593). Arcimboldo composed portraits of people out of pieces of fruit, sea animals or flowers. Yet in spite of his unusual artistic vision, *Flora* may be more authentic than *Mona Lisa*. Here's how I see it.

Alamy

Arcimboldo's vision is a good analogy for understanding the concept of the organism, which has undergone a dramatic transformation in recent decades. Symbiotic biology understands organisms in a similar way to Arcimboldo's paintings. Absolute individuals do not exist.

The Greek word *symbiosis* means 'living together'. Symbiotic biology is concerned with the construction of biological

Wikimedia Commons

systems through mergers. On this understanding, organisms are biological systems formed by the merger of simpler organisms into more complex ones. Each complex organism is seen as a symbiotic collective rather than an independent individual. Group selection principles apply, such that natural selection selects the collectives which together form fit organisms. By contrast, for earlier mainstream biology, natural selection has been about the selection of individual organisms. This approach treated organisms as undisputed individuals, and did not recognize the role of collectives in the formation or behaviour. It sees group selection as a problematic concept. From the perspective of such mainstream biology, Lisa Gherardini, the model for the *Mona Lisa*, is one organism. We are blind to the components, the hierarchies, the populations, that make up the ecological collective of her body. But Flora is more authentic than Mona Lisa because Arcimboldo's counterintuitive vision depicts the wholes as composites. Life is a set of transitional forms, or we might even say an organic composite form that is constantly changing. Evolution is the process of the change of the composites.

How would this work? Well, life began with microbes almost four billion years ago. Microbes built a giant system, a living planetary network of microbes, that altered planetary chemistry, for instance by oxygenating the atmosphere. Microbes also have their own precursor of biological communication. Eshel Ben-Jacob, physicist and microbiologist, was amongst the first to use the term 'bacterial linguistics' in such a context.

At least one billion years ago, the microbial planetary network produced the first symbiotic organisms, such as amoeba: eukaryotic cells, which are cells which have a nucleus. These types of cells, which include most of the cells in human bodies, are formed from the combination of two types of microbes – bacteria and archaea – when one type of cell swallowed the other, but the other stayed alive inside it.

Amoeba and her relatives, known as the protists, playing games with bacteria, learnt the crafts of construction with and communication amongst many cells. This merging of protist cells gradually produced multicelled organisms from once free-living individual cells. Microbes immediately joined in to help generate supersymbiotic collectives called holobionts, which means a complex individual organism living with all its individual bacteria, etc. All the organisms we see around us – humans, dogs, trees, fish – are holobionts. There are no plants or animals on the planet without associated microbiota. If you removed microbes from the biosphere, it would collapse. Holobionts themselves formed social communities – societies of insects, plants, people… In this way, some social communities turned into what have been called 'superorganisms'. The whole biosphere is a collage of diverse ecosystems: just like Arcimboldo's group portraits.

At the heart of symbiotic biology is the principle of biological coexistence. Whether we like it or not, we are microbial partners. Our bodies are ecological systems that comprise forty trillion cells, each one containing tens, hundreds or even thousands of integrated bacteria–mitochondria; and biologists estimate between forty and four hundred trillion more microbes which are not strictly speaking part of our bodies are present in our digestive tract and on our skin.

For symbiotic biology, life is a constantly changing and transitional complex form, and all organisms except the most basic microbes are chimeras – composite ecological systems thrown out into the biosphere by the thermodynamic storm of life, in the same sort of way that Arcimboldo painted human faces by merging seemingly incompatible images projected by the storm of his artistic mind onto his canvases.[21]

Agency (Purpose and Desire)

We now need to make sense of universal flux in the context of a world rich with living beings. Even though science is a powerful tool, we will never be able to identify all the organisms on Earth. Still, we can make educated guesses. The most recent estimate suggests that the total number of species – that is, all extant animals, plants, fungi, protists, bacteria and archaea – is between one and six billion.[22] Together, this amazing conglomerate of life forms that makes up the body of Gaia is estimated at 550 gigatons of carbon. How can we fit the truly gigantic Gaian body into the Heraclitean river and still preserve individuality in the great natural flow? To do so, we need a suitable theory of organisms. In other words, we need to populate the Heraclitean river with living things, and still remain within the bounds of science.

As we move forward in search of a suitable theory of organisms, a surprise awaits us. If you consult biology textbooks, you won't find any coherent theory of organisms. Yes, there are hints of such a theory, or parts of a theory at best, but not a unified and coherent whole. Why is it that we have the theory of evolution, but we don't have a theory of organisms? The answer to this intriguing question must have something to do with the peculiarities of academic biology. Most biologists think that when we ask a question such as 'What is an organism?' the answer already exists. If so, there is no need to ask the question in the first place. That is, there is no need to ask a question for which there is an obvious and definitive answer.

Let me formulate the seemingly definitive answer. According to the most dominant school of thought in biology today – the gene-centric view of life – organisms are biological machines fully determined by genes and the natural environment.[23] A good analogy is the meat grinder. The handle on the grinder is the genes. The metal body of the grinder is the natural environment. When the handle (the genes) and the body of the grinder (the natural environment) join forces, they start producing 'lumbering robots', in the words of Richard Dawkins. These are subdued and non-autonomous biological objects, or genetically programmed machines adapted to their environments. But we have already learnt that the concept of organisms as machines is troublesome and belongs to a realm of mechanistic biology from which we must escape.

One of the escape routes is an alternative theory, which views organisms as independent natural agents, underdetermined by genes and the environment.[24] Yes, genes and the environment play a role in how organisms work, but not a decisive one. What makes organisms from bacteria, archaea and protists to orchids, whales, elephants and humans free is their capacity to act independently. This capacity – called agency – is my second principle of life.[25] I also call it the 'purpose and desire' principle, for reasons I will explain.[26]

The notion of organisms as natural agents can be traced to the 18th-century German philosopher Immanuel Kant. Living creatures, according to Kant, are 'self-organising' entities – their parts act in unison to produce the whole and, vice versa, the whole exists for the sake of the parts. In other words, an organism is 'the cause and effect of itself'. The result of Kantian self-organisation is that the causal chain remains within organisms.

Scientists today call this phenomenon 'organisational closure'. Chilean biologists Humberto Maturana and Francisco Varela explained organisational closure through the concept of autopoiesis ('auto' meaning self; 'poiesis' meaning production, creation).[27] Autopoietic systems – all organisms from bacteria to mammals – contain sufficient processes within themselves to maintain the processual whole. This makes organisms autonomous and operationally closed agents. However, autopoietic systems remain structurally coupled with their environments. The result of organism–environment dynamics is the emergence of cognition. Organisms/agents are epistemic systems capable of understanding their surroundings. This makes organisms/agents natural learners or 'anticipatory systems' – a term coined by Robert Rosen in his remarkable 1985 book by the same name.[28]

Robert Rosen, who worked independently from Maturana and Varela, invented the mathematics of organisational closure through examining the traditional chain of causation that can be traced to Aristotle: material, formal, efficient and final causes. Rosen concluded that organisms are 'closed to efficient causation', which is a mathematical confirmation of autopoiesis.[29] This essentially means that organisms emerge spontaneously without the influence of any external causes – organisms make themselves; they are autopoietic systems. As the great French physiologist Claude Bernard said long ago: 'The constancy of the internal environment is the condition for a free and independent life'.[30] Bernard's 'constancy of the internal environment' is what agents do: strive to maintain their own bodies while interacting with the world. Rosen captured this striving of organisms by calling them anticipatory systems. Not only do organisms carry out and regulate their own organisation through exchanging matter, energy and information with their surroundings, they also produce internal models of themselves and their environment. Organisms make a difference to the world. Two young scientists, Maël Montévil and Matteo Mossio, have recently put the processual stamp on Rosen's mathematical model by combining processes and constraints to make Rosen's model more accessible.[31]

If we combine Maturana and Varela's autopoiesis and Rosen's anticipation, we can confidently say that there are no external causes that govern organisms. Instead, organisms are self-governed due to their organisational closure. This self-governance makes them purposive systems. Teleology is not an undesirable condition, as mechanistic biology suggests. The optimal natural habitat for organisms, as purposive natural agents, is the Heraclitean river. The purposiveness of organisms makes the river of life flow. Teleology – the capacity of organisms to extend actions into the future – moves the river of life forward into the territory of the unknown. And because organisms constantly deal with the unknown, they have the capacity to learn. This reinforces the notion that cognition is the essential feature of natural agents.

From the above, it follows that natural agents cannot exist without purpose and desire. This is not an endorsement of religious thinking, but an expression of the difficulties mechanistic biology faces. J.B.S. Haldane described teleology as a biologist's mistress, whom he can't live without but is ashamed to show in public. Pursuant to Aristotle, who had a sophisticated understanding of causality, Robert Rosen showed that the biologist should not be ashamed of the mistress. Organisms are part of the natural teleological flow. Even non-living complex systems – storms, autocatalytic chemical reactions, convection cells, etc. – show a naturalistic purpose of function. That is, they produce molecular chaos, or entropy, as they grow, 'in a teleological continuum of gradient reducing and energy spreading arrangements of cycling matter in regions of energy'.[32]

Eliminating purpose and desire from the science of life is equivalent to making life the opposite of what it is. Without purpose and desire, organisms resemble Monty Python's dead parrot – creatures 'bereft of life, they rest in peace', creatures that 'are no more' or 'have ceased to be'. Like the shopkeeper from the dead parrot sketch, played by Michael Palin, who stubbornly insists that the parrot is alive, 'pining for the fjords' or 'stunned', mainstream biology is blind to the purpose-and-desire principle and obsessed with the mechanised gene-centric version of life, despite – and this is the great paradox – the fact that mechanism is a feature of the non-living. We can sense the echoes of the dead parrot sketch in the words of J. Scott Turner: 'What modern Darwinism is asking us to admire is a husk of something once living, but with its vital core drained away as we have poked and prodded with our naughty thumbs until we are left with nothing but the beautiful shell. In short, the science of life has become disenchanted with life itself.'[33]

Symbiosis (Living Together)

The principle of agency allows us to generate a bottom-up view of the great natural flow. Organisms, or natural agents, make up the flow of life and are constantly pushing it into the territory of the unknown due to their purpose and desire. But what about the top-down view? Or, what's the 'glue' that holds natural agents and their environments together, maintaining a unified Heraclitean river of life?

The answer to this question is my third principle of life: symbiosis, or living together. Agents are not solitary creatures. They merge, and through mergers generate a unified flow – Gaia in its full glory. A heroine of modern biology, Lynn Margulis, showed us, against the tide of scientific opinion, that life is a constant merging of agents of different kinds, or symbiogenesis.[34] The basic and indivisible units of agency are the simplest prokaryotic cells, bacteria and archaea. All other natural agents – protists, fungi, plants and animals – emerge from them. Just look at us. We and all other animals are ecological collectives – composite agents – consisting of microbes and their mergers (see 'Flora or Mona Lisa?', page 32). Margulis described animals as 'coevolved microbial communities', permanently attached to the microbial cloud by virtue of our microbiomes.[35]

Indeed, symbiosis makes the biological world hierarchical: bacteria and archaea merge to produce protists, protists merge to produce plants and animals, these merge to produce ecological collectives, human culture emerges from ecological collectives, and all this is underpinned by the presence of microbes as the floating cloud enveloping the Earth, which holds the biosphere together.

Let us imagine that we can cut through the flow of the Heraclitean river Gaia, freeze the cut for a moment and then analyse the picture in detail. The cross-section of the flow will look like a huge painting covered with sectioned bodies of plants, animals, protists, fungi and their environments. If we then enlarge relevant areas of the picture, we will see that all those bodies and their environments are held together by the microbial glue of bacteria – the bacteriosphere.

The bacteriosphere – the ancient living planet, the precursor to modern Gaia – existed long before plants and animals. The bacteriosphere is the basis of the biosphere. The planetary bacterial system has been regulating the biogeochemical cycles of organic elements for the last three billion years. When James Lovelock and his friend William Golding used the name Gaia for the Earth's biosphere, the scientific community attacked the idea as non-scientific.[36] But Margulis saved Gaia from the dustbin of scientific history by showing us that microbes are at its base.[37] Natural agents, starting with microbes, form a biological supersystem, which in turn is capable of regulating itself through the process of homeorhesis. It is worth spending a few moments on this concept because it reveals to us a deep link to processual biology. Homeorhesis is the scientific view of the Heraclitean river of life.

A British geneticist, Conrad Hall Waddington, coined the term 'homeorhesis' around 1950.[38] He used it in the context of animal developmental processes. It means a steady flow. During the process of development, there are many disturbances to the developing biological system, but the system always returns to its developmental trajectory. The key word here is 'trajectory' – the system is moving from one state to another. Homeorhesis contrasts with an older term, 'homeostasis', which means a steady state. When Margulis and Lovelock thought about Gaia as a self-regulatory system, they initially used homeostasis and feedback loops to explain the system's dynamics. However, Margulis later realised that the reliance on feedback loops and the steady state contradicts the autopoietic nature of Gaia.[39] Autopoietic Gaia cannot be a static system. Instead, autopoietic Gaia is a moving system, a metaphoric river that constantly strives forward. This striving forward is a creative process that produces new and unexpected biological forms. Gaia is constantly reinventing itself.

Margulis also realised that the Gaian system differs from organisms/agents in its capacity to recycle internal waste.[40] All agents produce waste, ranging from oxygen produced by cyanobacteria to single-use plastic suffocating rivers and oceans. Organisms/agents are incapable of recycling their waste. (Some scientists think that we can invent methods for recycling our own ecological waste, but there are considerable obstacles, some of which betray overly optimistic technological hubris.) But Gaia does the business of recycling efficiently; the waste created by one type of agent becomes a necessary ingredient for the metabolic process of another.

The capacity of the Gaian system to move forward should be called homeorhesis, according to Margulis: the system regulates itself around moving set points, rather than the fixed set points of a steady state.[41] Temperatures, gas concentrations and other parameters of the Gaian system change with time. These changes are essential because Gaia is a dynamic system constantly moving into its next state. In other words, the system must have enough creativity and freedom to achieve the next state and remain a unified system, rather than be one that disintegrates and destroys itself. For this reason, Stuart Kauffman argued that the next state of the Gaian system is unpredictable, even by the system itself.[42] This is probably the first formal recognition that biology, unlike the mechanics of physics, has elements of freedom that may be likened to art (see also Chapters 1 and 9).

Additionally, the components of the Gaian system, connected through time by ancestry (genes as a form of biological memory, along with other forms of biological memory) and space by the changing appearances of matter, give us the full-blown Heraclitean river of life striving to expand its riverbed by moving into the territory of the unknown. Importantly, what keeps the flow of Gaia unified are not plants, animals, fungi or protists; the flow is held together by the bacterial network that has regulated biogeochemical cycles of organic elements for billions of years, and in doing so makes life possible.

Mind (Hyperthought)

The above three principles – universal flux, agency and symbiosis – may seem sufficient to describe Gaia as the Heraclitean river of life. Yet we can go one step further and ask the following question: if organisms are cognitive agents, and given that they are all symbiotically, and thus cognitively, joined into the unified flow of the Gaian river, is the Gaian system capable of decision-making? If the answer to this important question is affirmative, Gaia then fulfils Bateson's criteria of mind (see 'The Microbial Mind', page 14). Let's remember his first criterion: the mind is the unity of the parts that communicate with each other.[43] Indeed, the evidence that Gaia is a system capable of decision-making through the unity of its communicating parts is overwhelming. We know this from Lynn Margulis' and James Lovelock's detailed and well-articulated works, ranging from scientific articles to popular books.[44] Furthermore, Earth system science is converging on this view. Sergio Rubin and his colleagues have recently demonstrated Gaia's decision-making capacities. They have shown that Gaia may be critically dependent on feedforward elements of the system (future) produced by its autopoietic nature (past) (see 'Gaian Science versus Human Science', page 135).

Homeorhesis – the steady flow of the Gaian system – is indeed a mind-like process. Gaia integrates all cognitive agents that make it up into an intelligent whole greater than the sum of its parts. The important thing to realise is that the decision-making process in the Gaian system is not centralised. Instead, it is a distributed process requiring the participation of all agents. The distributed unity of agents gives birth to evolutionary novelties discovered through Gaia's constant striving into the territory of the unknown.

Thus, my final principle of life is mind in the Batesonian sense. It is worth repeating Bateson's maxim that served as the epigraph to Chapter 1: 'Mind is the essence of being alive.' Natural agents, irrespective of whether or not they have brains, are anticipatory systems that interact with each other in a mind-like way. These mind-like interactions seem to be widespread in nature. A recent model shows remarkable similarities between neuronal interactions in the brain and the cosmic network of galaxies, in spite of enormous differences in spatial scale.[45]

The second important thing to realise is that the intelligence of natural agents is not a form of cybernetic, or computer-like, intelligence. In the case of natural agents, mind represents more than computation.[46] The mind of natural agents is an ecological rather than a cybernetic phenomenon.

Let us elaborate on the distinction between the cybernetic or the computational metaphor for mind on one side, and its opposite, the environmental metaphor, on the other. There are two types of cognitive systems: artificial systems (intelligent machines or computers) and natural systems (organisms). While computers simulate cognitive actions, organisms live through the process of cognition. Cognition for computers is brought about externally by human programmers. Maturana and Varela call this feature of intelligent machines 'allopoiesis',[47] whereby machines make something different from themselves. On the other hand, cognition for organisms is part of their constitution. Organisms are self-made, through the process of autopoiesis, without the influence of external creators.[48]

How do cognitive systems work? To answer this question, we can rely on an elegant analysis by J. Scott Turner. He thought that cognitive systems do four broad things[49]:

1. Representation: Sensing environmental information and projecting it in the form of a cognitive map.
2. Tracking changes in the environment: Comparison between the cognitive map and sensory inputs.
3. Intentionality: Bringing the environment into conformity with the cognitive map of the world.
4. Creativity: Bringing the environment into conformity with unimagined cognitive worlds.

Intelligent machines can do three out of four cognitive actions. However, they are incapable of performing the final cognitive action: creativity. Organisms, on the other hand, are capable of performing all four cognitive actions. Thus, the key distinction between organisms and machines is the lack of creativity on the part of machines. Creativity is a feature of living systems.

This is a key reason why organisms, or natural agents, are autopoietic systems – those systems capable of bringing forth evolutionary novelty by inventing previously unimagined living forms. These living forms are new organisms and new environments, possibilities that will forever remain beyond mechanical systems. Machines remain within the realm of cybernetics – the man-made, artificial world of simulated cognition. Organisms, on the other hand, are ecological phenomena – the process of autopoiesis includes the emergence of not only new organisms, but also new environments created by these organisms. Robert Rosen and Aloisius Louie expressed the distinction between the cybernetic and ecological metaphors for mind in precise mathematical terms.[50]

When we view Gaia through the prism of the Batesonian mind, we can perhaps use the word 'hyperthought' to describe its creativity, or the capacity to invent new steps in its trajectory, new sections of the great natural flow.[51] Hyperthought describes the process of bringing together the cognitive actions of natural agents, in a distributed way, in the search for new and unimagined cognitive worlds. Gaia is like a sculptor making itself – the natural mind at work. This idea was first elaborated by a pre-Socratic philosopher, Anaxagoras. The Greek term he used to describe the natural mind at work was *Nous* (νοῦς).[52]

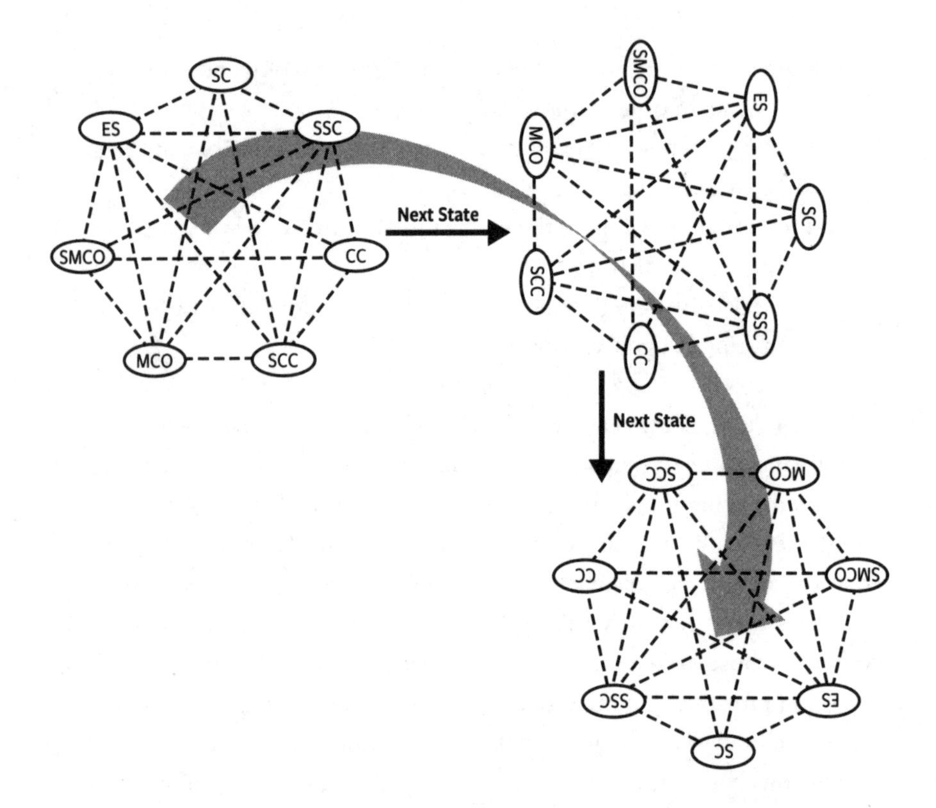

Figure 2.1. Life as the great natural flow. Gaia is depicted as the unity of the natural agents that make it up; the list of agents, from bacteria to ecosystems, is from Lynn Margulis' serial endosymbiosis theory.[53] The interconnectedness of agents represents mind or hyperthought. Homeorhesis represents the movement (flow) of the Gaian system from one state to another (grey arrow). Each state of the system is different, as symbolised by the rotation of the original system. SC – simple cells (bacteria and archaea); SSC – society of SCs (e.g. bacteriosphere); CC – composite cells (mergers of SCs; e.g. protists); SCC – societies of CCs (e.g. societies of amoebas); MCO – multicell organisms (plants and animals); SMCO – societies of MCOs (e.g. insect, plant or human societies); ES – ecological systems (e.g. mature forests as ecosystems consisting of microbes, protists, fungi, plants and animals).

Gregory Bateson used the term 'ecology of mind' to describe the totality of mind-like processes in nature.[54] If we combine Bateson's ideas with the process of homeorhesis, we get a picture of Gaia that I call 'physiology of mind' (see Figure 2.1). The term 'physiology' is an homage to the great French physiologist Claude Bernard, who is credited with explaining the process of homeostasis, or the steady state of biological systems. The concept of homeorhesis builds on Bernard's idea of homeostasis to allow for systems that flow around dynamic regulatory points, as in the case of Gaia. The Gaian system can be viewed as the network of natural agents that interact with each other in a mind-like manner. The system is not static; it is constantly moving, in the form of the great natural flow, around dynamic set points. The movement of the system creates new agents and new environments in a manner that remains beyond prediction by the system itself.

That nature is a system artistically creating itself – a kind of self-sculptor – we know from the controversial philosopher Martin Heidegger. He used the phrase '*Physis* is *poiesis*' to describe how nature works.[55] Gaia is more of an artist than a scientist. In the epic act of self-sculpting, a process Maturana and Varela called autopoiesis, Gaia cannot, in a truly artistic manner, predict its next state; we humans are a fleeting thought in the great natural mind of Gaia.

What Next?

Now we have a clear idea of the process of life. Life is not a thing or a mechanism. Life is a process of permanent change – the Heraclitean river Gaia striving into the territory of the unknown. The flow of Gaia is kept together by symbiosis, or the living together of natural agents. Gaia is also a form of mind in the Batesonian sense. In the mind-like search for the next section of the flow, all components of Gaia work together. This working together is a creative process. It results in the emergence of new organisms and their environments. The emergence of biological novelty is innovation in the artistic sense – even the Gaian system itself cannot predict its next state.

My vision of life differs from that depicted by mainstream, gene-centric biology. In the narrative of mainstream biology, new species (biological novelty) arise from the process of natural selection. Organisms have the capacity to produce more progeny than can survive. Thus, natural selection is the filter that allows only a fraction of progeny to enter the biological world. Mainstream biology gives unprecedented power to the filter; a fraction of this fraction of progeny will contain novel genetic mutations, making it sufficiently genetically different from the rest of the progeny (new species) and better suited to new environmental niches (adaptation). Thus, evolution, according to mainstream biology, is almost a mechanical process. Organisms evolve to fit already existing environments.

In my interpretation, natural selection is only an editor, but cannot be, under any circumstances, the creator.[56] Natural selection is a kind of a biological administrator that keeps record of which fraction of progeny, out of the maximum possible, survives. The administrator also takes account of which genes – both type and number – are propagated in the surviving fraction of the progeny. But there is no way that an act of biological administration can have creative powers.

Rather, the creation of biological novelty is a process far beyond the mechanics of gene-centrism. For example, in the greatest discontinuity in the history of life – the emergence of eukaryotes through the merger of bacteria and archaea – genes did not play a central role. The driver of evolution was the process of symbiosis, in which evolutionary novelty was the result of bacterial–archaeal combinatorics that extended beyond individual features of these organisms.[57] The innovation was a kind of phase transition in which the new phase of the system was qualitatively different from the previous one. The same can be said about the next big evolutionary transition: the emergence of multicellular organisms. This was another symbiosis-driven process.[58] And underpinning all evolutionary transitions are the creative and mind-like powers of the totality of the Gaian system. Gaia has existed, in less complex forms, ever since microbes turned the dead planet into a living one three billion years ago.

It is now time to set out on the next phase of our journey. The principles of life, now articulated, will help us move beyond the dogmas of anthropocentrism.

CHAPTER THREE

Pride and Prejudice

Man is the measure of all things.

PROTAGORAS

The goal of Chapter 3 is to lay the groundwork for understanding the concept of biocivilisations, with the help of principles introduced in Chapter 2. Chapter 3 is also the finale to Part 1, Beyond Humans, and a suitable introduction to Part 2, Brave New World, which will explore the science of life in an entirely new way, through the concept of biology as a civilising force.

Within this concept (see Chapter 4), humanity is virtually unimportant. This is because *Homo sapiens* is a young and evolutionarily inexperienced species. All other species we know of are our older evolutionary relatives. Insects, plants and amoebas are evolutionary long-distance runners that have been around for hundreds of millions of years,[1] not to mention bacteria and archaea, the true evolutionary marathon runners, masters of Earth for four billion years. In sharp contrast to all of them, humans have been around for a minuscule amount of evolutionary time: a few hundred thousand years.[2] This makes us short-distance runners with an uncertain future. The faster we sprint, the closer we get to an evolutionary exit.

Thus Protagoras' statement that 'Man is the measure of all things' is more complicated than it seems. Even though Protagoras did not know much about evolution, he was clever enough to sow the seed of relativity into our conception of truth – the more we know, the more we doubt. If we know little, our knowledge horizons are narrow. Equipped only with narrow epistemological horizons, we are likely to miss the big picture. As our knowledge expands, not only do we start seeing glimpses of the big picture, but we also begin to doubt our naïve world view.

Belief in the special status of 'man' – the would-be 'highest life form' – expressed in the quasi-religious *Homo Deus* or the popular Anthropocene,

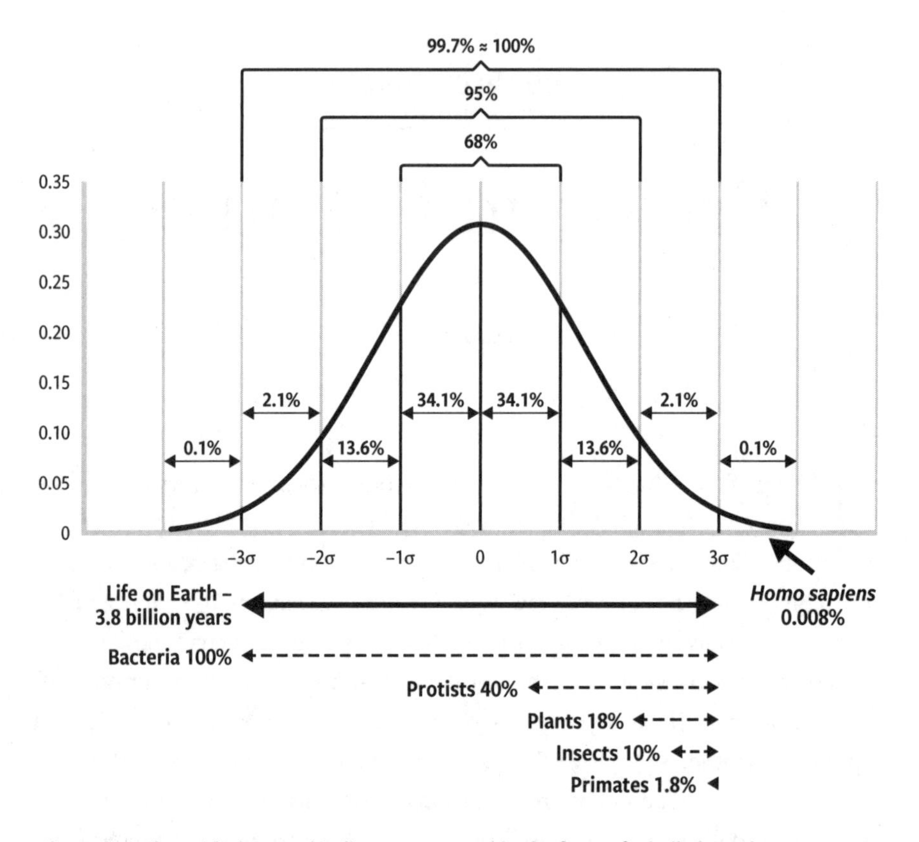

Figure 3.1. The evolutionary timeline represented in the form of a bell-shaped curve (normal distribution) and the three-sigma (σ) rule (aka the 68–95–99.7 rule). Timelines of bacteria, protists, plants, insects, primates and *Homo sapiens* are indicated as percentages of the bacterial lifetime.

which places humanity above nature, is flawed because it ignores the big picture of natural history. When viewed on the evolutionary scale, the timeline of *Homo sapiens* borders on the insignificant, as you can see in Figure 3.1.

Figure 3.1 is a visual representation of timelines of life forms in the course of evolution; the longer the timeline of a species, the more typical it is as a component of the biosphere. The figure has the form of a bell-shaped Gaussian curve, representing normal distribution. It is combined with a useful heuristic in empirical sciences known as the three-sigma rule or the 68–95–99.7 rule.[3] One sigma (σ) represents one standard deviation from the mean. According to the three-sigma rule, nearly all values are located within three standard deviations of the mean. This means that the 99.7%

probability is equivalent to 100%. Any value located outside three sigmas is a non-representative outlier.

The evolutionary reality is crystal clear. While other species, from bacteria to insects, have made clear imprints on the curve of life, the position of *Homo sapiens* is equivalent to a non-representative outlier because we have only been around for 0.008% of evolutionary time. Again, we are short-distance runners with an uncertain future, a tiny part of the Gaian river that may or may not survive the next turn. Yet we are obsessed with the self-deification that has persisted in European culture from St Paul and Martin Luther to Hegel and modern-day transhumanists.[4]

The value of Figure 3.1 is not only in evolutionary realism; it allows us to radicalise Protagoras' statement. Given our evolutionary insignificance, humanity as a godly species becomes the greatest self-delusion. Only with post-delusional self-correction, tantamount to the Copernican turn in biology (see Chapters 1 and 11), can we 'measure all things' in a more realistic way. The concept of biocivilisations is exactly that. To understand the living world, we must confront the delusional streak in our nature. This doesn't prevent us from being proud of our achievements, but pride is sweeter when we separate it from our prejudices.

Prejudice No. 1: Intelligence

The principles of agency and mind, which I introduced in Chapter 2, acknowledge that all organisms are natural agents with cognitive capacities; bacteria are no less smart than human beings. These capacities enable natural agents to integrate into the Gaian mind-like system. The interplay between natural agents and the Gaian system is fluid, or processual. The principle that everything flows reveals that Gaia self-regulates through the steady flow of the system, or homeorhesis. Life is an intelligent, mind-like process.

The trouble starts when we humans impose our concept of intelligence on the rest of life. The idea of human superiority can be traced to the first biologist, Aristotle. In his *History of Animals*, Aristotle argued that plants are inferior beings.[5] Aristotle's work served as the basis for the medieval concept *Scala Naturae*, or the Great Chain of Being. The anthropomorphised God and angels were placed at the top of the natural hierarchy, followed by humans as God-like creatures. The rest of the scale was occupied by lowly constituents, including animals, plants and minerals. The great systematiser Carl Linnaeus did not challenge *Systema Naturae*. He was a religious man who thought that he had been put on Earth by God to do the job of biological classification; his motto was *Deus creavit, Linnaeus disposuit* ('God created, Linnaeus classified'). Even though he placed humans in the kingdom of animals, he thought *Homo sapiens* a godly species. Things changed with Lamarck and Darwin in the 19th century. *Scala Naturae* was replaced by the concept of evolution by natural selection and a tree of life without a god.

But there was one big difference between Lamarck and Darwin. Darwin expelled the mind from biology, whereas Lamarck tried to retain it. According to Bateson, Lamarck was right and Darwin wrong. In the essay 'Pathologies of Epistemology', from his book *Steps to an Ecology of Mind*, Bateson outlined a vision of biology remarkably close to the concept of biocivilisations (see 'Organism plus Environment', page 8). In brief, Bateson thought, contrary to Darwin, that the unit of evolutionary survival is neither the family line nor the species. The unit of survival, according to Bateson, is the unit of mind.[6]

The unit of mind carries within itself the cognitive relationship between the organism and its environment. The key implication of this statement is that cognition is a biological universal. Organisms, from bacteria to elephants, are lifelong learners. The learning process consists of the abilities to sense stimuli coming from the environment, process the stimuli or 'think them over' to create knowledge, and then direct behaviour based on acquired knowledge by actively changing the environment. It is important to stress that the process of learning is not cybernetic but ecological. Mind is more than computation.

How would the unit of mind work? The best analogy is Arthur Koestler's concept of the holon.[7] At the heart of the concept is the Janus effect. The Roman god Janus has two faces, one facing left and one facing right. If we were to turn Janus' head through 90 degrees, we could see the face looking upward as the face of a subordinate part, and the face looking downward as the face of a self-contained whole. Koestler called this dichotomous double-faced unit a holon (see Figure 3.2 A). A holon is neither the whole nor the part, but both the part and the whole at the same time. The concept of a holon is extremely close to the processual nature of life. Arcimboldo's *Flora* is the epitome of holarchy: the merging of holons to generate larger semi-autonomous systems (see 'Flora or Mona Lisa?', page 32), which remain dichotomous with respect to the rest of nature.

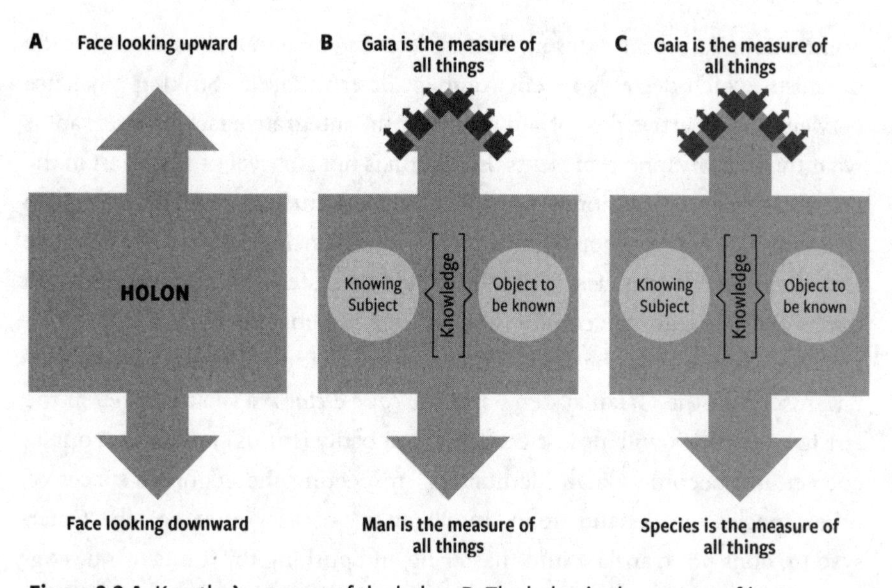

Figure 3.2 A. Koestler's concept of the holon. **B.** The holon in the context of human knowledge. **C.** The holon as an evolutionary unit of mind.

We can use the holon to explain the unit of mind (see Figure 3.2 B). Humans, and all other living beings, are knowing subjects. The world – or more precisely, the local environment we inhabit – is the object we need to understand. In other words, the object looks like an existential puzzle to the subject. To solve the existential puzzle, the subject generates empirical knowledge about the object. This includes changing the object according to the subject's existential needs. The components of civilisations we've built – from agriculture, electricity and aeroplanes to the internet and sprawling modern cities – are evidence that we are masters of our cognitive space. Humanity is indeed the measure of all things. In the context of the holon, the mastery of our cognitive space is the face looking downward on the world; the assertive face of a self-contained whole.

However, the face of the master must become the face of the servant. The face looking upward means that we must integrate into the Gaian system. This integration is not easy. We suddenly face a different kind of existential puzzle: how to integrate our cognitive space into the holarchy of cognitive spaces developed by other species, and by doing so remain part of the Gaian system. It is now clear that humans are in the process of learning this integrative lesson the hard way. Environmental alarms sound over our future, representing a growing awareness that accepting subservience is as important as claiming domination. To repeat Bateson: 'We are learning by bitter experience that the organism which destroys its environment destroys itself.' Striking a balance between the assertive face of a master and the integrative face of a servant is what the unit of mind represents. Evolution is not survival of the fittest in the Darwinian sense; evolution is survival of the most intelligent in the Lamarckian or Batesonian sense, whereby intelligence is a balancing act between assertive and integrative tendencies. Gaia, the mind-like system, is the measure of all things once we leave the comfort zone of our cognitive space.

If we assume only the face of the master, we forever remain an authoritarian sore on the Gaian system, and the *coup d'état* we've attempted in the last few centuries will almost certainly end badly (for us) unless we course-correct and become Gaian 'democrats', respecting the cognitive spaces of other species we depend upon. Intelligence, in the context of the Gaian system, does not translate into mastering and pushing the limits of our cognitive capacities through mechanised science. Intelligence means finding a balance between the two faces of our dichotomous nature.

We can now universalise the unit of mind (Figure 3.2 C). Along with agency and mind, cognition is a biological universal. Every species has its own cognitive space that relies on senses and experiences to acquire knowledge about the world. The cognitive space of a tick is as functional as the cognitive space of a human.[8] Every species is the master of its own cognitive space. But this mastery has a price. The self-assertiveness of species-specific cognitive spaces must be balanced by the integrative capacity required to become part of the Gaian system. Only those species intelligent enough to balance natural assertiveness with an integrative capacity to enter the Gaian 'democracy' can leave an imprint on the curve of life (see Figure 3.1, page 48), or even remain part of the Gaian river for any length of time. But no species, apart from prokaryotic microbes, is a permanent member of the Gaian system. Species come and go, but bacteria are forever.

The evolutionary unit of mind is a challenge to mainstream biology primarily because biology professors, with a few exceptions, do not recognise cognition as a biological universal. The unit of mind is a non-starter in the world of academia; there is no place for it in the heavily mechanised science of life. But if there is truth to the notion that the unit of evolutionary survival is equivalent to the unit of mind – as opposed to the unit of the fittest biological machine controlled by genes – we are faced with an existential paradox. Instead of helping us solve the environmental crisis, mainstream biology unwittingly exacerbates it by defending the mechanised science of life.

Indeed, most academic biologists recognise higher cognitive functions, including intelligence and mind, only in organisms that have brains. Neuroscientists and their heavily mechanised arsenal of molecular biology insist that intelligence correlates with brain complexity.[9] This is a step back to Aristotle and *Scala Naturae*. We humans have the most complex brains, and thus we occupy the prime position in the natural hierarchy of intelligence.

Table 3.1. Biomass of the biosphere

Taxon	Biomass in gigatons of carbon
Viruses	0.2
Bacteria	70
Archaea	7
Fungi	12
Protists	4
Plants	450
Animals	2
	Total ~550

This anthropocentric position is buttressed by a long line of philosophers and scientists who have grappled with the 'hard problem of consciousness' as the evidence that *Homo sapiens* have the most superior minds in nature.[10] Lynn Margulis and Dorion Sagan challenged this anthropocentric interpretation of consciousness in their book *What Is Life?*: 'Not just animals are conscious, but every organic being, every autopoietic cell is conscious. In the simplest sense consciousness is an awareness of the outside world.'[11]

As I mentioned in Chapter 2, the biomass of Gaia is 550 gigatons of carbon. This biomass is intelligent because it consists of natural agents integrated into the Gaian body in a mind-like fashion. The interactions between agents and the system are how Bateson described them. The evolutionary unit of survival is equivalent to the unit of mind. The distribution of units of mind that form the body of Gaia is shown in Table 3.1.[12]

The units of mind that dominate the Gaian body are plants and bacteria. Together they comprise 520 gigatons of carbon, or 95% of the Gaian body. Animals, the only beings that have brains, constitute less than 1% of the Gaian body. Gaia and its constituents are intelligent, despite the presence of brains only in traces.

But academic biologists are still under the influence of Aristotle and *Scala Naturae*. They think that organisms without brains lack intelligence and that Gaia is 99% stupid. Ernst Mayr endorsed this heavily anthropocentric interpretation of intelligence in his dialogue with Carl Sagan:

> *Sagan adopts the principle that 'it is better to be smart than to be stupid', but life on Earth refutes this claim. Amongst all the forms of life, neither prokaryotes nor protists, fungi nor plants have evolved intelligence, as they should have if they were 'better'. In the more than twenty-eight phyla of animals, intelligence evolved in only one (chordates) and doubtfully also in cephalopods. In the thousands of subdivisions of the chordates, high intelligence developed in only one, primates, and even there only in one small subdivision. So much for the putative inevitability of the development of high intelligence because 'it is better to be smart'.*[13]

Mayr's line of thinking is indirectly endorsed by AI researchers who argue that everything below superintelligence or the Machinocene – anticipated to arrive by the mid-21st century (this is one of the delusions of the mechanised view of the world) – must be inherently stupid.[14] AI researchers proclaimed the death of biology through the concept of post-biological evolution; machines will run the world in a not-so-distant future (see 'Prejudice No. 4: Post-Biological Evolution', page 61). Biology is simply a springboard for a higher form of mechanical mind.

Mayr and these AI researchers are wrong. Mayr was wrong because he was promoting the zoo-centrism of modern synthesis, which is biased towards less than 1% of life forms.[15] The reality conveyed by Figure 3.1 (page 48) is a devastating blow to the Mayrian and neo-Darwinian view of the living world. Brains exist in nature only in traces (see Table 3.1). The AI researchers are wrong because they are biased towards a simulated version of life that will remain just that – a simulated version existing only in human models of reality. Mind and mechanism do not go together. The attitude of these AI researchers is best summarised by the expression 'brain chauvinism', coined by the champion of plant intelligence, Anthony Trewavas.[16] His colleague James Shapiro, the champion of bacterial intelligence, is even more critical of anthropic arrogance: 'Bacteria are far more sophisticated than human beings at controlling complex operations.'[17]

Prejudice No. 2: Anthropocene

God and angels were removed from the top of the natural hierarchy in the 19th century, and the position of man was downgraded to the kingdom of animals by Lamarck and Darwin. The scientific story behind evolution by natural selection rightly challenged the perceived dominance of *Homo sapiens* over other forms of life. Yet, as we entered the 21st century, there was a sudden revival of the spirit of *Scala Naturae*, this time without God and angels at the top. Instead, the cultural hubris endorsed by a large number of scientists placed humans at the top of the hierarchy once occupied by God. Proponents of the Anthropocene seem to endorse a secular version of *Scala Naturae*: 'Humans have replaced nature as the dominant environmental force on Earth.'[18]

But how can a non-representative outlier on the curve of life (see Figure 3.1, page 48), which constitutes only one-hundredth of a single percent of life by weight (see Table 3.1), be the master of all planetary life, an all-powerful controller of the Gaian river?

The antidote to this delusion is the principles of life. In the context of 'everything flows', *Homo sapiens* represent an insignificant stream in the Gaian river – so insignificant that the river will easily neutralise any disturbance caused by such a tiny stream. Agency implies that cognition is a biological universal, far from the anthropocentric position, as seen through Mayr's spectacles, that non-human organisms are outright stupid. The principle of symbiosis, or living together, represents a form of Gaian democracy whereby all species are equal, but the Anthropocene casts us as more of an unmanageable bull in a china shop. Terms like 'biological annihilation' and 'Necrocene' may betray our clumsiness, but they also reveal the arrogance caused by a lack of awareness of the symbiotic nature of life. Finally, the principle of mind implies that there can be no mechanical solution to the environmental crisis. This is because Gaia, as a mind-like system, is homeorhetic; it requires the creativity of its constituents and of itself, which is not compatible with the one-dimensional outlook of mechanised science.

I hope that by now the stakes of this dialectic are becoming clear, including the potential futility of human engineering of the biosphere aimed at resolving the anthropogenic environmental crisis. In some circles, this engineering programme is called Earth stewardship and the assumption is that the 'intelligent' part of the biosphere – aka the noosphere – is capable of altering the biosphere in line with its own needs.[19]

For example, the biosphere-engineering programme of sorts known as the Paris Agreement (which almost all countries signed) is specific in requiring each country to limit the rise of global temperature to below 2°C relative to preindustrial levels, as well as make greater efforts to limit the rise to below 1.5°C.[20] The main method for reducing the temperature is to lower emissions of greenhouse gases, primarily CO_2, into the atmosphere. Each country must increase its ambitions every five years. Collective progress will be monitored by a global stocktake, with the first assessment expected in 2023. The goal is to achieve the complete decarbonisation of emissions between 2040 and 2060. This will require establishing new reservoirs for CO_2 absorption, known as carbon sinks.

The expression 'Earth stewardship', however, betrays our ineptitude concerning understanding nature. Lynn Margulis was ruthless in her criticism: 'We humans aren't stewards of anything except our flimsy ships, but we are inordinately arrogant...especially scientists.'[21] Indeed, scientists, who enjoy immense trust from politicians and the public, are largely concerned with implementing engineering solutions consistent with human dominance over nature. Humanity's collective ego is so big that some astrophysicists, for example, regard the planet as a 'spaceship Earth', controlled by scientists as privileged representatives of humanity.[22]

This attitude – no questions asked – is anti-scientific (see Richard Feynman's comments on humanist pride, page 63), and it may lead to an even more destructive trajectory. Our ineptitude with regard to the living world will void our ticket for long-term membership in the Gaian democracy. Living systems are not computable – they cannot be engineered by science – because they are autopoietic.[23] Indeed, computation, which engineering relies upon, may be limited to the point of uselessness as a solution to the problems of life because life is not a cybernetic phenomenon.[24]

The autopoietic nature of Gaia is democratic; there are no privileged species in Gaian democracy. All species are equal. Perturbations caused by one species are balanced by the Gaia-generated structural changes aimed at preserving its autopoietic nature. These Gaia-generated structural changes, as a response to perturbations, are multiple, distributed and cannot be predicted by mechanical input–output cybernetic models.[25]

The real captain on spaceship Earth is Gaia. Humans are like a group of drunken sailors who have decided to rebel against the captain. Our models are commensurate with the short-term effect of our drunken state. Models last only until Gaia responds with a series of distributed and unpredictable changes that will wake us up from our self-aggrandising dreams.

Is there any way out of this impasse? My answer is that practitioners of the mainstream science of life – a thought collective that subscribes to the notion that life is a machine – should pay attention to the works of a small group of their colleagues who do not view biology as a mechanised science subservient to physics.[26] This small group includes Lynn Margulis, Gregory Bateson, Francisco Varela, Humberto Maturana and Robert Rosen. Collectively, their works argue that nature is a creative system that cannot be tamed by computation-based mechanics and humanised engineering.

Prejudice No. 3: Sentience and Consciousness

Sentience is usually described as the capacity of organisms to sense environmental stimuli. Sensations may involve positive or negative experiences: pain, the pleasure of tasting food, positive movement towards light (phototaxis), etc. Consciousness, on the other hand, is the capacity to have subjective experiences or awareness of environmental stimuli. The differences between sentience and consciousness are small. Ethologists argue that all sentient beings are conscious. However, not all conscious beings are sentient. For example, there could be forms of damage to the body of an organism that do not produce any sensations.

Science is divided about which beings are considered sentient and conscious. As things stand, mainstream science remains Aristotelian about the hierarchy of mental phenomena. At the top of the hierarchy is human mentality; the mentality of all other beings is either inferior to this or non-existent. René Descartes argued that animals are passive automata, with no sense of pain or emotions. This line of thinking directly contributed to mind–body dualism, a position that is still influential amongst philosophers of mind.

Science has advanced since Descartes to the point that sentience is recognised in at least some animals. For example, the British parliament introduced the Animal Welfare (Sentience) Bill in early 2021, partly as a way to combat animal cruelty. (Great Britain has a reasonably good track record with this, at least when it comes to domesticated animals. The first legislation of this kind in the world, the Cruel Treatment of Cattle Act, was passed in 1822, followed by the Protection of Animals Act in 1911.) The Animal Welfare (Sentience) Bill, now Act, is summarised in three lines on the government's website[27]:

- Government introduces Bill to formally recognise animals as sentient beings
- Animal Sentience Committee will put animal sentience at the heart of government policy
- Bill was introduced as part of the government's first-of-a-kind Action Plan for Animal Welfare

The Act is explicit, however, about which animals – exclusively vertebrates – are considered sentient. While we cannot expect a government to protect the lives of insects in the same way it protects the lives of cattle or dogs, it doesn't change the fact that all organisms, including insects, are sentient and conscious beings. Science is gradually converging on this point. For example, the idea that insects such as ants are passive automata is clearly incorrect. Scientists have learnt that ants are capable of subjective feelings, and even have personalities.[28] Granted, we must remain cautious when interpreting the mental worlds of others, as we do not know them from the inside, but even simple awareness that someone or something else is conscious and sentient may help expand our epistemological horizons.

The most significant problem for mainstream science is the 'brain barrier'. Many scientists are only willing to consider sentience and consciousness in organisms that have brains and nervous systems. Plants and microbes are brainless, so they are typically not considered sentient or conscious. This Aristotelian position, however, has been vigorously challenged both recently and historically. Botanists Anthony Trewavas, František Baluška and their colleagues argue that the brain and nervous system are not required for consciousness.[29] Plants, according to them, have sentience and consciousness, despite having different physiologies from animals. Indeed, Charles Darwin argued that the equivalent of a plant brain is their root system.[30]

The 19th-century French physiologist Claude Bernard proposed response to anaesthetics as a litmus test. Any organism that responds to anaesthesia, he argued, passes the sentience test. More recently, Trewavas and Baluška have discovered that if you treat plants with anaesthetics, the plants can no longer respond to environmental stimuli. Plants also synthesise anaesthetic compounds in response to stress and wounds.[31]

Furthermore, Trewavas and Baluška argue that sentience and consciousness started with the first forms of life: prokaryotic cells such as bacteria.[32]

All other life forms, including plants and animals, inherited sentience and consciousness from the first cells. The key question they're trying to answer is: what are the molecular properties of cells that make them sentient? These properties can be examined in plants to confirm or refute plant sentience.

Trewavas and Baluška propose that three molecular properties of cells contribute to sentience: excitable cell membranes, molecular components of the cytoskeleton and the flexible behaviour of proteins.[33] The cell membrane is the cell's window onto the world – a dynamic lipid bilayer with quasi-crystalline properties. This window is sensitive to stimuli, or information, coming from the external world. In other words, it is responsive to the environment. This causes the cell to become 'aware' of itself and its environment. When plants are treated with anaesthetics, their cell membranes lose their sensitivity, making the plants insensitive or 'blind' to the outside world.

The cytoskeleton is a cell's internal structure that determines the cell's shape and facilitates cell division. It is present, in different forms, in all cells, including prokaryotic (bacteria and archaea) as well as eukaryotic (protists, fungi, plants and animals) cells. The cytoskeleton is made up of molecular complexes known as microtubules and actin filaments. These structures are chemically excitable and can oscillate in response to stimuli. Like cell membranes, microtubules and actin filaments are affected by anaesthetics.

Here, Trewavas and Baluška have encountered unexpected allies. Mathematician and physicist Roger Penrose and anaesthesiologist Stuart Hameroff argue that the microtubules in brain cells are responsible for human consciousness.[34] Penrose thinks that the actual cause of consciousness results from changes in the quantum states of the microtubules. By extension, it's possible that the excitability of cell membranes is also due to changes in the quantum states of these structures.

Finally, proteins show remarkable spatial flexibility in their three-dimensional folding, independent of the genetic code; this might also contribute to consciousness. For example, proteins select one of several possible shapes in response to physicochemical conditions within cells. In particular, transmembrane proteins, which embed into cell membranes, behave as cellular Maxwell's demons. In thermodynamics, Maxwell's demon is a thought experiment devised by James Clerk Maxwell. The imaginary demon controls a tiny door between two gas chambers. The demon can 'see' components of the gas: molecules and atoms. The demon then quickly opens the door and lets fast-moving molecules and atoms exclusively into one chamber. Since the temperature of a gas depends on the speed of movement of its constituent molecules and atoms, one chamber will become heated and the other will cool down. This violates the second law of thermodynamics by decreasing entropy without any work. Transmembrane proteins, through their different configurations, may act as cellular gatekeepers, helping cells decrease entropy, but without violating the second law of thermodynamics. This non-violability is captured in a term coined by Dorion Sagan, 'Maxwellian angel'; organismhood, or autopoiesis, means the ability to self-produce by dissipating energy while it lasts.[35] The second law of thermodynamics is unbreakable. Even non-living complex systems, from storms to weather patterns, behave as Maxwellian angels (see Chapter 4).

The next question is how cellular sentience and consciousness lead to a larger-order supracellular sentience and consciousness in plants and animals. This question, according to Trewavas and Baluška, is relatively easy to answer using the principle of symbiosis. Fifty years on, Margulis' account of endosymbiotic eukaryogenesis in the form of bacterial–archaeal mergers still stands. However, the order of events in the mergers and the nature of host–symbiont relationships has changed in light of new phylogenetic evidence.[36]

Thus, eukaryotic cells are already ecological communities made up of multiple cells. This means that sentience and consciousness above prokaryotic cells are ecological phenomena – the distinction between organisms and environment is blurred (see 'Flora or Mona Lisa?', page 32). Similarly, mergers of eukaryotic cells to produce multicellular organisms are ecological processes; the bodies of plants and animals are ecological collectives of cells. These ecological collectives of cells are at the same time corporate bodies with their own sense of sentience and consciousness. In animals, the brain and nervous system become centres of sentience that constantly assess ever-changing organism–environment interactions.[37] In plants, which are sessile organisms, there is no need for the central nervous system. Instead, plant sentience relies on different structures such as phloem and root systems.[38]

In summary, sentience and consciousness are features of all organisms. This is contrary to the view of mainstream science, which recognises sentience only in some animals. Without sentience and consciousness, there can be no cognition and intelligence. According to Roger Penrose, cognition and intelligence represent the part of consciousness responsible for understanding. Sentience, on the other hand, represents the part of consciousness responsible for feelings, emotions and a subjective sense of oneself. And, to my larger point, in the words of Roger Penrose: 'Whatever consciousness is, it is not computation. Or, it is not a physical process which can be described by computation.'[39]

Prejudice No. 4: Post-Biological Evolution

Post-biological evolution, which is championed by some astronomers, AI researchers and philosophers, is a scenario that places anthropocentrism in the driver's seat of cosmic evolution.[40] In the context of the principles of life, as well as in light of the information presented in Figure 3.1 (page 48), post-biological evolution is another prejudice cooked up in the kitchen of a mechanistic interpretation of life.

The main argument for post-biological evolution is that the emergence of humanity signals a major shift in cosmic evolution. Biological evolution dominated by biomolecules, Darwinian natural selection and random mutations has been supplanted by human cultural evolution. This shift accelerates a cosmic transformation of matter and moves it in a particular direction: relentless improvement of the human species by medical interventions and integration with AI, resulting in 'cosmic humanity' – or, in other words, masters of the universe. These new masters of the universe will travel freely through the cosmos, might become immortal, will communicate with extraterrestrials and will be capable of doing things beyond our wildest imagination.[41] Post-biological evolution has been heavily influenced by an interpretation of life in the style of modern synthesis (which is heavily zoo-centric).

However, the edifice of post-biological evolution, as a scientific or philosophical concept (which its proponents now call the 'mainstream'), lacks solid foundations and is easy to refute.[42] Any serious scientific concept must be built on some form of universalism, a wide perspective that exposes fallacies of exclusivism. For example, life is easily distinguished from non-life by analytic chemistry. If you have two random samples collected from unknown locations on planet Earth, one of which contains DNA and proteins (sample 1) and another which contains some calcium salts (sample 2), it's a no-brainer which sample is likely to be organic – a high probability for sample 1 and a low probability for sample 2. Carbon, oxygen and hydrogen in DNA and proteins make up over 90% of organic matter. Calcium, on the other hand, is a minor component of organic matter found in the bones of some animals and the shells of foraminifera.

Post-biological evolution uses a form of logic that favours the argument that sample 2 is likely to be composed of organic matter. This is an exclusivist argument. Indeed, calcium is found in some life forms, but only as a minority compound. The exclusivist argument rests on an exceedingly low probability – an outlier elbows its way into the centre (see Figure 3.1, page 48). The consequence is a flood of fallacies, a lack of realism and the emergence of secular religion – that is, science based on faith. Indeed, some critics of AI argue that its proponents rely on religious arguments.[43]

Why is it that post-biological evolution is based on false logic? The answer is simple. The interpretation of the living world that proponents of post-biological evolution rely on is almost exclusively based on modern synthesis and neo-Darwinism. Given that modern synthesis is biased towards <1% of life forms, it is not surprising that post-biological evolution, as a scientific or philosophical concept, is heavily biased towards exceedingly small probabilities.

For example, many scientists think *Homo sapiens* invented civilisation and is the only species to have ever demonstrated evidence of culture, including language, mind and memory (see Chapter 1). Biocivilisation is an argument against this human exclusivism. There is not a single species in the history of life on Earth that has not possessed the communication skills necessary for language. Likewise, there is not a single species without collective memory or the practice of natural learning, which leads to the development of technologies. Insects developed their technologies, such as agriculture, millions of years before us (see Chapters 6 and 10), and the bacterial biotechnology known as photosynthesis has fuelled life on Earth for the last three billion years (see Chapter 4).

The exclusivist argument makes us blind to the biocivilisations built by bacteria, protists, insects, plants and other species over the course of the evolutionary timeline that preceded humanity (99.99% of it). Some achievements of bacteria will forever remain beyond human technological capabilities.[44] Mathematical solutions that honeybees achieved millions of years ago – the honeybee algorithm – are now used to solve problems of human internet traffic (see Chapter 9). Many species rely on aesthetics for procreation – a possibility that turns the evolutionary principle of the survival of the fittest favoured by modern synthesis on its head (see Chapter 9). Plants have their own culture that becomes apparent in the complex ecologies of mature forests (see Chapter 6). There is nothing in the repertoire of human civilisation that is biologically or culturally unique. Every single feature of civilisation that we think of as invented by *Homo sapiens* existed, in one form or another, millions or billions of years before us. Gaian democracy is the democracy of biocivilisations.

The final arguments against post-biological evolution are autopoiesis and relational biology. Life is an ecological phenomenon in which organisms and environments merge into units of mind. Post-biological evolution is the forced mechanisation of life – taming life with the help of a cybernetic or computational metaphor to turn it into a mechanical mind dominated by AI-mediated superintelligence. However, life and mechanism do not go together (see Chapter 2). Life and mechanism belong to different categories of logic.

Humanist Pride

It is now time to turn from prejudices to pride. Some of the most brilliant human minds contributed to the development and progression of ideas that have made the information and cosmic ages possible. So many scientists, including Enrico Fermi, Frank Drake and Carl Sagan, contributed brilliant ideas to the development of astrobiology. Similarly, Alan Turing, Norbert Weiner and Marvin Minsky made the science of informatics possible. The ingenuity of these researchers is a testament to the strength of the human investigative spirit.

However, it was almost inevitable that humans would overreach in our investigative quests and enter a territory of anthropic arrogance. Scientists err, but good science is self-correcting. Richard Feynman, as was often the case, said it best: 'Each generation that discovers something from its experience must pass that on, but it must pass that on with a delicate balance of respect and disrespect...to teach both to accept and reject the past with a kind of balance that takes considerable skill. Science alone of all the subjects contains within itself the lesson of the danger of belief in the infallibility of the greatest teachers of the preceding generation.'[45]

In other words, the pride of human achievement is sweeter when we separate it from prejudices. Indeed, we can interpret science as an endless process of separating pride from prejudice. We should be proud of Sir Isaac Newton's achievements in celestial mechanics because they helped us understand the world, but we should also be aware of his prejudices, such as his obsession with alchemy.[46]

By the same token, we should be proud of what AI research has brought to human technological progress (computers, the internet, social media, etc.). At the same time, we should not be scared to criticise those aspects of AI that have been exaggerated. One of those prejudices is that human intelligence is so superior to the rest of life that its simulated version will run the world in the near future.[47]

To summarise, life is a mind-like process that requires all participants to constantly learn. The process of learning started 3.8 billion years ago here on Earth and it lasts until the present day. The evidence for this is us – we are the most recent product of natural learning. Without microbial learning, there would be no humans, plants or animals. The integral part of the process of learning is understanding. Whatever this understanding may be – from an understanding of bacteria, protists and plants, to the human understanding of the world – it is not computational. Mind is more than computation. Roger Penrose singled out the following forms of non-computational understanding: mathematical understanding, human understanding generally, human consciousness, animal consciousness and life.[48] Penrose encourages physicists to drop the computational burden from their scientific shoulders and go beyond provisional theories such as quantum mechanics.[49]

With inspiration from Penrose, I believe that the current science of life is a provisional theory – provisional because it is based on mechanism. The concept of biocivilisations is my attempt to go beyond this provisional theory. We need to replace anthropocentric prejudices with the humanist pride of Shapiro's position described in Chapter 1 – understanding humanity's true place in the world.

Part II

BRAVE NEW WORLD

CHAPTER FOUR

Civilising Force

Everything was fermenting, growing, rising with the magic yeast of life. The joy of living, like a gentle wind, swept in a broad surge indiscriminately through fields and towns, through walls and fences, through wood and flesh.

BORIS PASTERNAK, *Doctor Zhivago*[1]

Life is a river that started flowing four billion years ago here on Earth. There may be rivers of life on other planets too. But before we start searching for extraterrestrial rivers of life, we must understand our own in its full glory. This is where the concept of biocivilisations becomes central. In Part 1, I presented the concept in broad terms (Chapter 1), argued against the mechanistic interpretation of life by outlining the principles that make biocivilisations understandable (Chapter 2) and identified areas where the mainstream science of life differs from the outlook of biocivilisations (Chapter 3).

Part 2, Brave New World, is not an homage to Aldous Huxley and his futuristic novel so much as an invitation to step into a brilliant new reality that will be revealed to us, if only we are courageous enough to remove ourselves from the centre of it. Only by being willing to dial back the anthropocentrism that characterises (and perhaps even defines) our species can we begin to see the wisdom of the parallel worlds that surround us – from the peculiar civilisation of bacteria that made the planet alive, to majestic civilisations built by insects such as ants, termites and honeybees. The bravery here is an act of breaking long-held dogmas. Science not only helped us build modern human civilisation but also revealed the paradox of existence – from the famous quantum mechanics–based thought experiment known as Schrödinger's cat, we know that the nature of reality may not be what we think it is. Anthropocentric reality must be deconstructed: we have to biocivilise the world to understand it.

The first thing to note in the brave new world of biocivilisations is that life is a force. This makes biology independent from physics. In physics, there are forces, fields and interactions. For example, physicists have identified four fundamental forces or interactions: strong, weak, electromagnetic and gravitational. In the Standard Model, weak, strong and electromagnetic interactions act on fundamental particles.[2] Many physicists aspire to integrate these three forces into a Grand Unified Theory. Beyond that are attempts – such as string theory and M-theory – to formulate a Theory of Everything, an ultimate theory that will unify all forces and explain all aspects of the universe.[3] Some of the greatest scientific minds the world has ever seen, including Albert Einstein, have failed to describe a theory that would unify all the forces in the physical world.

But biology does not have the unification problem that drives physicists into unpleasant battles.[4] In biology, all forces are unified into one: life. The key argument of this book is that life is a *civilising* force – hence *biocivilisations*. In this chapter, I will explain how that civilising force works. In subsequent chapters (5 to 10) it will become clear how the force of life transforms into the civilising actions of all species, not just *Homo sapiens*.

The Magic Yeast of Life

Poets and artists are often more effective than scientists at capturing the essence of phenomena that surround us (see the epigraph to this chapter). However, 'the magic yeast of life' and 'the joy of living, like a gentle wind' require further elaboration to turn such a poetic vision into a proper scientific one. A 'gentle wind' was a storm in the evolutionary past, and may become one again.

The river and the energy flux go hand in hand. Our river Gaia emerged from the wave of energy, solar or non-solar, and it has been riding on solar energy for billions of years. The amount of energy transferred from the Sun to the Earth – the enormous radiant flux – is estimated at 128 petawatts or 128,000 terawatts daily.[5] Most of that energy fuels weather, climate and ocean circulation. However, a small part, 90 terawatts, is captured by life processes.[6]

The energy flowing through an open thermodynamic system such as planet Earth (or any other planet exposed to a similar wave of energy coming from a Sun equivalent) imposes orderliness on the system. 'Orderliness' is the keyword for this chapter because it is synonymous with the civilising action of life which, when applied to evolution as a whole, turns into the sweeping beauty of the living world that I call biocivilisations: species-specific cognitive spaces that cause orderliness of breathtaking proportions (see Chapters 5 to 10 for details). In short, biocivilisations represent the endless orderliness of life, from the mystery of the bacterial language in which the story of evolution is told, to the artistry of sand paintings produced by male pufferfish to attract females, to the wisdom of mature forests in which microbes, fungi, plants and animals are integrated into the most complex form of orderliness on Earth.[7] What keeps all biocivilisations together is Gaia: the four principles – everything flows, agency, symbiosis and mind – unified into the powerful river of life.

Most importantly, there is a clear scientific explanation behind biocivilisations: orderliness is the consequence of planetary-scale thermodynamic processes; there is nothing magical about it. The 'magic yeast of life' is a poetic appreciation of the orderliness of life. What poets or artists see as magic, however, is not necessarily contrary to science, even though many scientists bristle at poetic or artistic expressions of reality as contrary rather than complementary.[8] Indeed, these two approaches, scientific and artistic, are most effective when they are sufficiently delineated from one another to protect their respective contributions to collective understanding, but sufficiently proximate that they can speak the same language in a constructive exchange of views. I find Pasternak's description of life superior to dry (and often erroneous) scientific descriptions of life. At the same time, I also find Pasternak's description to be naïve, as I think will become clear below.

What is the science behind the orderliness of life and ultimately behind biocivilisations? The answer is not complicated at all. Wherever there is energy, that energy will carry out work. The work performed by useful energy results in changes to a system that look like orderly changes. Typical examples of orderliness on the planetary scale are weather patterns that we see not only here on Earth, but also on Mars, Venus and Jupiter. A good example is Jupiter's Great Red Spot, a giant Earth-sized cyclone that has existed for at least 350 years. In fact, Jupiter itself has a series of gorgeous bands of clouds that move in orderly ways because of the energy emitted from the planet's interior. Other cosmic examples of orderliness include interstellar clouds known as nebulae, which have names that reflect our interpretations of their incredible patterns: the Pillars of Creation, the Crab Nebula, the Horsehead Nebula and the Cat's Eye Nebula (a cosmic zoology, of sorts, far, far above our heads, in the sky). Indeed, the entire universe resembles an orderly cosmic web of galaxies. Scientists who have analysed the pattern behind the cosmic web of galaxies have concluded that the cosmic-scale pattern is similar to a much smaller pattern: the network of nerve cells that form our brains.[9] This pattern of brain cells has been repeated a total of 107 billion times during the embryonic development of all humans who have ever lived.[10]

The similarity in the forms of orderliness generated by living (brain) and non-living (web of galaxies) processes leads us to the central scientific point behind biocivilisations. One fragment of the planetary-scale orderliness is life. The second law of thermodynamics describes the natural function of energy spread using the concept of entropy. In the process of energy spreading, work occurs that facilitates planetary weather patterns and autopoiesis, but the cumulative effect is more energy spreading. Thus, 'the inherently telic tendency of energy to spread informs life, whose living systems measurably produce more entropy, that is, reduce more gradients and delocalize more concentrated sources of energy than would be the case without them'.[11] This means that Gaia emerged from the orderliness of the Earth's thermodynamic system, then rode on that orderliness and amplified it by producing the finest forms of orderliness in the history of the cosmos: life in the form of biocivilisations.

Translated into the language of science, life is a process that goes beyond physics (see Chapter 2). Life processes channel energy into their organisation (where channelling equals thermodynamic work or constraining energy into a few degrees of freedom) and by doing so produce autopoietic units or holons (cells to Gaia) that perpetuate themselves as long as the energy is available, and in the process create (1) entropy that leads to thermodynamic equilibrium slower than would occur without them and (2) unpredictable patterns that go beyond the mechanics of classical physics. In the 19th century, many scientists were tempted to associate the inherent unpredictability of life with a mysterious force called *elan vital*. However, the life force as we understand it today is non-mysterious. Stuart Kauffman stated this with clarity: 'We may have found the "life force": not a nonphysical mystery but a marvel and a different mystery of unprestatable becoming.'[12]

We can now return to Pasternak's poetic vision and correct it. Indeed, it is easy to fall into the trap of idealising life as a gentle wind. Standing at the top of a meadow on a summer day and observing the tranquil unity of plants and animals, bees and flowers, ants and hills, and listening to the birds' songs and the mooing sounds of cows, gives us an impression of an idyllic quality of nature.

However, the gentle wind of nature is deceptive. It happens only when the river Gaia enters wide and slow passages that allow all members of bio-civilisations to enjoy a temporary tranquillity and thrive within it. But if we go upstream along Gaia's flow – in science this is known as predicting the past, or retrodiction – we can discover that Gaia was extremely wild in the evolutionary past; for example, the Oxygen Catastrophe, caused by cyano-bacteria producing oxygen approximately two billion years ago, killed many anaerobic bacteria. The reality is that the enormous force behind the river of life makes all members of biocivilisations temporary 'riders on the storm', as in the memorable song by Jim Morrison and The Doors. We ride on the powerful stream of the cosmic solar energy tamed for us by the Gaian system.

Evolutionary Retrodiction

How can we link orderliness – a widespread natural phenomenon visible in non-living and living worlds, from Jupiter's bands of clouds to human brains – with the heavily anthropocentric concept of civilisation? To answer this question, I shall use a technique called evolutionary retrodiction. In science, retrodiction means investigating the past using state-of-the-art techniques to get a more complete picture of reality. The evolutionary past, when properly understood, may serve as a corrector of the present and a suitable guide to projecting the future.

Here is how evolutionary retrodiction works in the present context. First, we identify phenomena usually associated with human civilisation. These might include science, art, architecture, engineering, medicine, mathematics, agriculture and other attributes of our modern civilised practices and behaviour. Next, we look at how these forms of civilised behaviour influence human culture. Typical products of civilised modernity include sprawling cities, beautifully engineered cars, aeroplanes and spacecraft; modern hospitals and medical schools in which powerful new therapies are invented, from manipulating genes to harnessing cells for regenerative medicine; artistic exhibitions, from wrapping bridges to extravagant forms of performance art; scientific discoveries that will help us conquer space; genetically engineered food that will make conventional agriculture obsolete; and many others.

We then analyse whether examples of civilised behaviour typical of human culture existed in the evolutionary past amongst microbes, plants and animals. I shall use this technique extensively in Chapters 5 to 10. The technique is similar to analogy-making as a form of reasoning. An American polymath, Douglas R. Hofstadter, described analogy-making as 'the very blue that fills the whole sky of cognition'.[13]

To give you a flavour of what's coming in the following chapters, here are two examples of evolutionary retrodiction. According to historian Niall Ferguson, human civilisations revolve around cities.[14] Take modern cities such as New York, Dubai, London, Shanghai, Moscow and Singapore. A close look at their organisational patterns reveals extensive similarities – they look like replicas of each other. Tall skyscrapers enclosed in glass, streets full of cars, underground transportation systems, airports at the outskirts, hospitals, police stations, shopping centres, cinemas, theatres, restaurants and cafes, museums, gyms, schools and universities. Polities that never sleep.

But when we delve deep into the biology behind human cities, we discover surprises. Cities are not a human invention at all. Social insects – in particular, ants and termites – invented cities as forms of social organisation millions of years before us. Take underground settlements built by leafcutter ants from the genera *Atta* and *Acromyrmex*: there is no substantive difference between an ant city and a human city. Scientists who researched and excavated cities made by leafcutter ants discovered equivalents of transportation systems, specialised buildings, food distribution centres, police stations, military barracks and even funeral parlours. What unifies ants and humans is a biological phenomenon known as eusociality, the result of which is the social organisation that gives birth to cities (see 'Eusociality and the Origin of Cities', opposite).

Cities, which emerged as a form of social organisation amongst insects, re-emerged millions of years later in human civilisation. This means that patterns of orderliness repeat throughout evolution. Mainstream biology calls this phenomenon convergent evolution. In brief, convergent evolution, also sometimes called homoplasy, refers to similar traits developed by

Eusociality and the Origin of Cities

If you put all humans living on the planet into an imaginary tin like sardines, the tin would be 2 kilometres long, wide and high. Amazingly, all the ants in the world would fill a similar-sized tin. Yet, despite their huge numbers, insects such as ants manage to thrive without overwhelming the natural world.

Insects are true inventors of technology. They have been using their technologies for over fifty million years in perfect ecological balance. By contrast, humans are amateurish technologists from an environmental perspective. Our technologies put the existence of our species and the entire biosphere at risk. By studying the peculiar social world of insects and their technology, we can learn how to live in greater biological harmony with the planet.

One thing humans already share with several species of insects is a practice known as 'eusociality'. This is the highest form of social behaviour. It involves a sophisticated division of labour, with different generations working together and different individuals carrying out different jobs, including giving birth and raising children. The most notable eusocialists on the planet are rare species of insects (ants, termites and bees) and *Homo sapiens*.

Our eusociality is what enabled us – and insects – to develop technology. Members of a eusocial species will identify and carry out the task they are most suited to, whether that's guarding the nest or seeking out food, while avoiding the tasks of other members. In this way, a group of animals will spontaneously organise itself into a collective we call a 'superorganism' that can generate technology. The first evolutionary form of technology was agriculture, which was actually invented by ants and termites fifty million years before us. Agriculture can be defined as the technological process of producing food on a large scale. For example, leafcutter ants turn green leaf biomass into food using their impressive gardening skills, formidable technical skills and symbiosis with a fungus.

The human practice of agriculture started 10,000 years ago. It is no more than a copy of the gardening and technical skills developed by insects, but in our case, agriculture enabled the creation of economic surplus. This in turn helped the emergence of written language, literature, mathematics, philosophy, art and eventually science. So the real roots of our technological abilities are in agriculture.

Biological Civilisations

Eusociality and agriculture enabled the social conquest of Earth, first by insects and then by us. The first global civilisation was the civilisation of insects, or the 'Insectocene'. By investigating various forms of the Insectocene, we can learn valuable lessons for our technological future. The meaning of the term 'civilisation' revolves around cities. The likes of Rome, London and New York have all been seen as the centres of ancient or modern human civilisations. All eusocial insects practising agriculture have their own cities, and you may be surprised to learn that they do not lag behind human cities in their technical sophistication.

Take leafcutter ants again. Their metropolises are probably the most intricate structures ever built underground. They have huge gardens at their centres, connected by excellent highways. Other structures scattered around include rubbish collection depots, food distribution centres, army barracks and police stations. There are even funeral service stations. Some ant cities are huge. If ants were as big as humans, the settlement of *Formica yessensis* ants on the island of Hokkaido in Japan would be much larger than Tokyo. Similarly, if termites were human-sized, then the height of an average termite mound in Africa would be the same as the tallest human construction, the Burj Khalifa in Dubai.

Other insect cities are examples of aesthetic sophistication. For instance, cities built by honeybees from the material excreted by their bodies are real examples of natural beauty. We need to start looking at insects as our older and more experienced eusocial relatives. These urbanised animals and their sophisticated social behaviour may contain important clues for making our technological future safer. We should be brave enough to enter their world without fear.

Somewhere in the peculiar social world of insects hides the secret of the ecological balance so absent from our world. We have to find this soon: the secret of how to integrate our technologies with the ecological system inhabited by at least nine million species. Otherwise, it may be too late.[15]

evolutionarily unrelated species. For example, flight emerged independently in insects, birds, dinosaurs, fish and bats. But I prefer a different term: 'biological periodicity'. According to Antonio Lima-de-Faria, biological periodicity is displayed in the functions and forms of living organisms.[16] It occurs repeatedly and regularly and is not linked to organismal complexity or the influence of the environment. The first form of vision in evolution was bacterial. A spherical cell of a cyanobacterial species behaved as a lens and represented a primitive eye.[17] Vision is not an exclusive characteristic of animal bodies – yes, it has emerged many times in animals, but it has also emerged in plants.[18] Patterns of orderliness, from flight and vision to cities and culture, flow through the river of life and emerge periodically and unexpectedly, but with persistent regularity.

A second example of evolutionary retrodiction is art. Art is considered one of the highest forms of orderliness ever produced by human civilisation. Kenneth Clark, in the famous TV series *Civilisation: A Personal View*, broadcast on the BBC in 1969, equated the art produced during the European Renaissance with European civilisation itself.

But if you read Darwin's *The Descent of Man, and Selection in Relation to Sex*, you will find evidence that survival of the fittest through natural selection is only one part of the evolutionary game; the other part is sexual selection. In his 2017 book *The Evolution of Beauty*, Richard O. Prum describes how sexual selection works.[19] It is an artistic game in which the males of fish and bird species are real artists who prepare love nests to attract females. Females, on the other hand, are excellent art critics. They visit many love nests. Then they go back to the ones that seem most appealing to them. They take a close look at the artist and his creation. Is the ceiling of the nest boring, or does it radiate the originality of the Sistine Chapel? Does the love nest possess the mystique of the Taj Mahal, or does it resemble an ordinary palace? Are there traces of Picasso in the arrangement of colourful objects that make up the interior, or is it just a bad knockoff? Only when all the artistic pieces fall together in her head does she give a clear signal to the chosen male partner. In other words, art is not a human invention (see Chapter 9); animals were creating art millions of years before the Renaissance artists who so impressed Kenneth Clark.

These two examples of evolutionary retrodiction (cities and art) show that the term 'civilisation' is inaccurate in its reference to human exclusivism (see Chapter 3). The French invented the term and used it as an antonym for the barbarism that pervaded the Dark Ages of European culture until the Renaissance. The English writer Samuel Johnson objected to the term and instead advanced 'civility' to mean polite urban behaviour.[20] However, both the French and English words, civilisation and civility, ignore the fact that ants and termites – not humans – were the first urbanised animals. Human civilisation is the tip of the iceberg; all forms of civilised behaviour, from art to medicine to engineering, can be identified in cultures developed by other species, as you will see in the coming chapters. This does not reduce the value of human civilisation. A more complete understanding of the evolutionary past places modern human civilisation in proper evolutionary context, and – if we are wise – may even help us successfully chart our course into the future.

To mitigate the human exclusivism in the term 'civilisation', I have added to it the prefix 'bio'. The result is a new term, 'biocivilisations', that describes the orderliness running through the river of life and all its constituents, from microbes to human beings. In my interpretation, life is a civilising force. Biocivilisations of bacteria or ants are not inferior to various forms of human biocivilisations. If we accept the new interpretation, the problem of human exclusivism disappears. Orderliness is a biological universal. There is no reason to single out the orderliness associated with human civilisation and give it a God-like status at the expense of the rest of the living world. By insisting on human exclusivism, we risk ending up a one-species autocracy that will violently collapse as Gaia shakes us off to preserve the autopoietic balance of the biosphere. In the river of life, all species are equal. The orderliness that runs through it is a form of biological democracy.

What Is Life?

The above question – the central question in the science of biology – has been the subject of many books, including Erwin Schrödinger's book by the same name. The mathematician Freeman Dyson praised Schrödinger for directing biologists to investigate DNA and genes as the basis of life, but argued that the narrow perspective of gene-centrism may be misleading. In the summary of this chapter, I present several premises behind the concept of biocivilisations which challenge mainstream biology's understanding of life.

- We can understand life only if we temper the application of mechanistic principles, appropriate for physics, to biology. Physics can be helpful in understanding biology, but only to a limited extent. The basic principles of life – everything flows, agency, symbiosis and mind – recognise the independence of biology from physics.
- Life started as a planetary-scale, top-down process and is driven by the Sun–Earth energetic axis. The key phenomenon regulating life, as the planetary-scale force, is homeorhesis. The best metaphor for life is the Gaian river. Gaia relies on its autopoietic mind to regulate its flow and take it into new and unpredictable territories.
- The river of life is a civilising force, best understood as orderliness running through the river in the manner of biological periodicity. The orderliness of life is the continuation of orderliness emanating from the orderliness of planetary-scale thermodynamic processes.
- Separating forms of orderliness created by humans – modern civilisation – from the orderliness created by other life forms, and granting humanity God-like status (irrespective of whether that status is religious or secular), is scientifically and philosophically wrong. There is no human civilisation; there are only various human biocivilisations,

and they are only a strand in the web of biocivilisations created by millions of other species.
- The intelligence of biocivilisations is judged by Gaia, not by us. Cows, lilies or amoebas may look stupid to us when we apply our own standards, but these become irrelevant when the Gaian river selects life forms suitable for long-term inclusion in its flow. Likewise, the persistence of non-living, gradient-reducing systems, such as the Great Red Spot of Jupiter, may be a sign of intelligence that goes beyond human understanding.
- Gaia's hyperthought is the king of the biosphere (see Figure 2.1, page 44). The evolutionary unit of survival is the unit of mind (see Figure 3.2, page 51).
- The new biology of biocivilisations requires a broader basis that must include scientific ideas developed by anti-mechanistic biologists, and aesthetic elements to nature that can only be understood through art (see Chapter 9).

Modern humans should be proud of our achievements in medicine, agriculture, art, science and engineering, but we didn't invent them. Understanding the achievements of other species in these fields may help us understand the true position of humanity in the web of the Gaian mind. In Chapters 5 to 10, I will explore parallel worlds ignored by mechanistic biology. We will learn about the wisdom of non-human communicators (Chapter 5), engineers (Chapter 6), scientists (Chapter 7), doctors (Chapter 8), artists (Chapter 9) and farmers (Chapter 10). The picture I draft is only a scratch on the surface of the Gaian mind.

CHAPTER FIVE

Communicators

Thus, we regard as regrettable the conventional concatenation of Darwin's name with evolution, because other modalities must also be considered.

NIGEL GOLDENFELD and CARL WOESE[1]

An educated person living anywhere in the world today is accustomed to the idea that *Homo sapiens* is the most intelligent species in the history of life. This form of anthropocentrism gives us carte blanche to design the everyday world that reflects our perceived superiority over nature. A walk through Walmart, Sainsbury's or Tesco in anglophone countries, Aldi or Lidl in Germany, or Carrefour in France reveals how we shape that world. Sections dedicated to our animal friends – cats and dogs – contain beautifully designed cans and boxes of food filled with menus richer than those of small village pubs. Here are some options from Tesco's cat food section: gourmet gold pâté with ocean fish, gold savoury cake with chicken, salmon and trout stick and classic terrine chicken.[2]

A walk through the same shops reveals how we treat enemies: ant-killer powder; ant and cockroach killer; fly and wasp killer; fly, wasp and mosquito killer; rat and mouse killer – not to mention various bacteria killers in the form of soaps, sanitisers, wipes and sprays.[3] This simple example reveals the dual face of humanity in the treatment of the natural world.

While most of us can tolerate such a cognitively dissonant relationship to the natural world, our sense of superiority as the arbiters of it is not without problems.[4] This is the mentality, after all, that leads us into the future. There is little difference between religious and secular outlooks when it comes to predicting how human intelligence will shape the future of the natural world. For example, it was a Catholic priest, Pierre Teilhard de Chardin, who predicted the Omega Point, a point in the future when the cognitive layer of the biosphere (i.e. the noosphere, or humanity) will enable the unification of the entire universe in the style of Christian Logos.[5] But it was a mathematician and physicist named Frank Tipler who similarly argued that intelligent life – that is, human intelligence – will take over matter (as the key constituent of reality) in the future, leading to a singularity similar to the Omega Point.[6]

Scientific futurism is only slightly different. Futurists argue that *Homo sapiens* – or more precisely, human culture – has surpassed evolution.[7] It took evolution 3.8 billion years to create *Homo sapiens*, arguably the most intelligent species in the history of life. But such intelligent humans need incomparably less time to create intelligent machines. According to futurist Ray Kurzweil and philosopher Nick Bostrom, intelligent machines may surpass our cognitive abilities in only a few decades, leading to a technological singularity or superintelligence.[8] Kurzweil predicts that superintelligence will turn planet Earth into a gigantic computer by the year 2099.[9] These arguments ignore the fact that the bacteriosphere has been running Earth's biogeochemical affairs for billions of years, with a kind of intelligence that will forever remain beyond human capacities.[10]

Challenging Anthropocentrism

The brave new world of biocivilisations challenges our prejudices by turning anthropocentrism on its head. The visual representation of the turn is presented in Figure 5.1.

The box expected to be occupied by human civilisation or human intelligence is replaced with the five kingdoms of life. I deliberately use the biological classification based on five kingdoms, rather than the more recent three-domain classification invented by Carl Woese.[11] The five-kingdom classification – in particular the version proposed by Lynn Margulis and Karlene V. Schwartz (which is based on an earlier version created by Robert Whittaker) – provides a richer background for the concept of biocivilisations because Margulis and Schwartz incorporated the prokaryote–eukaryote distinction, which is essential for the principle of symbiosis.[12] Also, bacteria represent one kingdom, divided into two subkingdoms, archaea and eubacteria, which eliminates the hassle of making frequent distinctions about subtle differences between archaea and bacteria, as is necessary with Woese's three-domain classification. (If anywhere in the book I ignore archaea at the

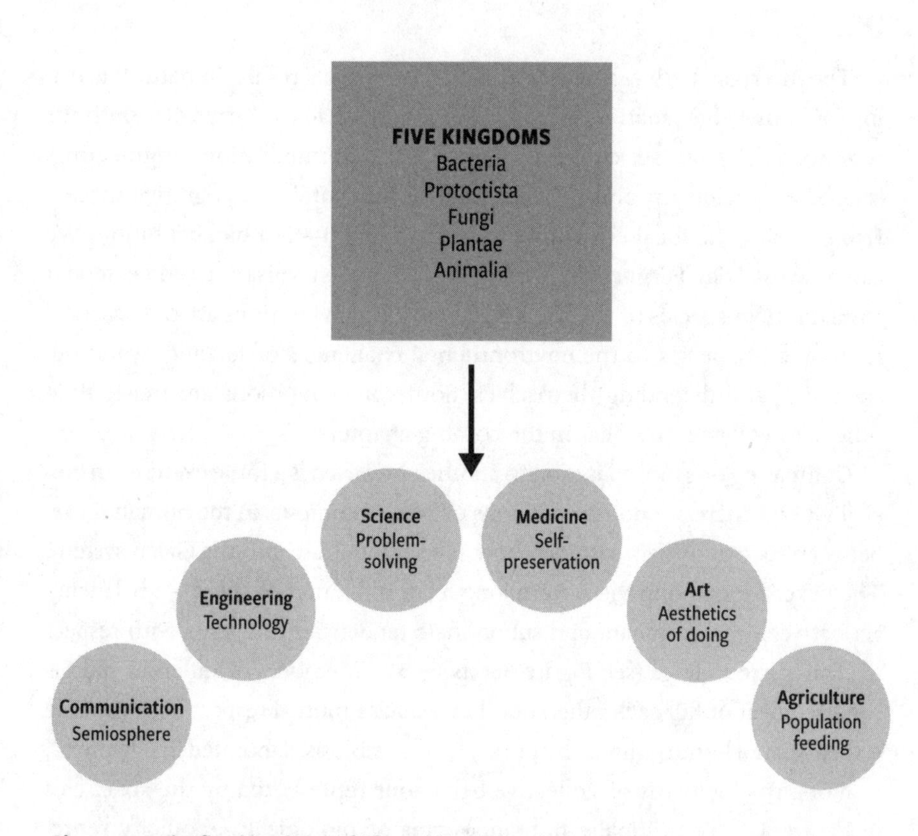

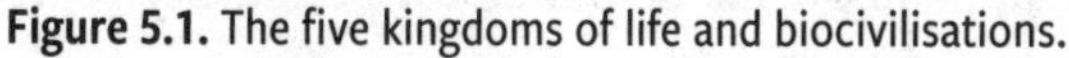
Figure 5.1. The five kingdoms of life and biocivilisations.

expense of bacteria, this is the justification.) Finally, Margulis' and Schwartz's classification includes two superkingdoms: Prokarya (bacteria) and Eukarya (the remaining four kingdoms). This brings the five-kingdom classification closer to Woese's three-domain classification and also highlights the dominance of prokaryotic microbes in the evolution of life on Earth.

In Figure 5.1, *Homo sapiens* is like any other species in the multimillion-part conglomerate of species. Each species is a group of agents that share the perceptual world, or cognitive space, typical of that species. Organisms, or agents, sense local environments and change them using effector organs. For example, honeybees sense flowers in order to collect nectar and turn the nectar into honey. In the process, honeybees invent a methodology for optimal nectar collection from a given flower patch, and by doing so they satisfy their own food needs, service the plants' reproductive needs and allow other animals to share the food they produce.

The interplay between sensor and effector organs results in natural learning, or knowledge creation, which in some cases leads to forms of behaviour represented by the six circles in Figure 5.1: communication, engineering, science, medicine, art and agriculture. The sum total of a cognitive space – from sensing the local environment to changing it – is a biocivilisation. We can now use Niall Ferguson's description of human civilisations and extend it to all member species of the five kingdoms.[13] Biocivilisations are the practical responses of species to the environmental challenges of feeding, watering, sheltering and defending themselves. Some biocivilisations are richer than others, as will become clear in the coming chapters.

Comparing one biocivilisation to another, however, is a futile exercise. In biocivilisations, there are no competitions or beauty contests, in the human sense, between species. What matters is how species integrate into the Gaian system. The ticket for entering the system long-term is the unit of mind – a balancing act between the dominant and subordinate tendencies of agents with respect to their surroundings (see Figure 3.2, page 51). Members of different species can eat, cheat or kill each other, but they exercise mutual respect because Gaia is their shared home – this is the principle of symbiosis elaborated in Chapter 2.

None of the forms of collective behaviour represented by the six circles in Figure 5.1 are originally human. Forms of biological periodicity represented by the six circles emerged in the distant evolutionary past. In the case of human culture, these forms of behaviour are replicas of ur-art, ur-communication or ur-science. At the same time, each form of civilised behaviour has certain species-specific features that remain unique. For example, art and engineering have uniquely human features, while medicine and science have certain unique features found in bacterial or ant biocivilisations, as you will see in the coming chapters.

The Living Theatre

After this necessary introduction, we can start to explore each of the six circles from Figure 5.1. In this chapter, we will focus on the first circle: communication. The closest synonym for communication is language. It is virtually impossible to find a species composed of individual organisms that do not communicate with each other. After painstaking work that took several decades, microbiologists discovered multiple bacterial languages ranging from the chemical language known as quorum sensing, to electrical signal exchange similar to neuronal signalling in the brain.[14] Given that bacteria are everywhere – high in the sky, deep underground and in virtually every single terrestrial habitat – this means that the planet is reverberating with bacterial laughing, shouting, chatting and perhaps even storytelling (see 'The Internet of Living Things', page 16); we just can't necessarily understand and translate these into human-language equivalents of words, sentences and complex meaning.

Plants also talk to each other. Almost every single part of a plant's body, from its roots to its leaves, secretes volatile organic compounds (VOCs).[15] VOCs act as messages that spread by diffusion in the air or soil. When the messages reach other plants, the plants react as we might upon hearing a specific word or sentence. Scientists have identified thousands of different VOCs that serve as words or sentences in plant language. These include hormones such as ethylene and jasmonate, as well as monoterpenes that plants release when insects damage them, all acting as words. Combinatorics of words constitute sentences that plants 'utter' in response to various situations ranging from defence to attracting herbivores. The plant body is the equivalent of many human mouths – the end of our vocal tract through which words and sentences are transmitted. This perhaps reflects the fact that plants lack a brain – an organ that, in animals, coordinates cognitive functions and directs conscious communication through one channel, the only vocal tract we have. By the same token, plants have more than two ears. They listen to messages with their entire body.

Understanding bacterial and plant biocivilisations requires special skills that mainstream scientists often lack. Stephen Jay Gould, in his review of Edward O. Wilson's book *The Diversity of Life*, reminded Wilson, an expert in sociobiology (a discipline exclusively concerned with animals), 'that we live in the Age of Bacteria (as it was in the beginning, is now and ever shall be, until the world ends)'.[16] Giving animals precedence over other life forms is an Aristotelian atavism that Gould called 'arrogance' and 'exaggerated self-importance'.

Because of this exaggerated self-importance, investigation into bacterial and plant biocivilisations has mostly been confined to scientific fringes, maverick scientists and disciplines with a small number of followers. In Chapter 3, I mentioned a group of botanists led by Anthony Trewavas and František Baluška, whose ideas on plant intelligence are disputed by their colleagues. Scientific disciplines like biosemiotics, evolutionary epistemology, relational biology and the systems view of life are almost entirely unknown to the public and even to scientists, who rely on a limited number of high-impact scientific journals, such as *Nature* and *Science*, to form an opinion about a research topic.[17]

These fringe disciplines challenge long-held dogmas. A particularly stubborn dogma is the perceived superiority of human communication. However, research shows that syntax and semantics (basic elements of human languages) are also common in bacterial and plant communication.[18] Languages as a form of intraspecies communication are widespread.

One challenge is that the concept of language does not capture all of the elements of natural communication systems. For example, interspecies communication is also widespread. Scientists call this cross-kingdom communication.[19] When plants release VOCs, the recipients are not only other plants. Bacteria, viruses, protists, fungi, insects and nematodes are all versed in cross-kingdom communication.[20] This is remarkable because no human being can understand plants without using complicated and indirect procedures from the scientific method, and even then incompletely. But some species are born with the ability to communicate with another kingdom of life. There is an old saying attributed to Charles V, Holy Roman Emperor: *Quot linguas calles, tot homines vales* ('As many languages as you speak, so many individuals are you worth'). Charles V boasted that he spoke French with men, Italian with women, Spanish with God and German with his horse. However, certain organisms, including bacteria, plants and insects, surpass every human individual in their communication skills; they can talk to organisms in other kingdoms of life.[21]

Cross-kingdom communication might explain why some insects, like ants, seem to live in harmony with trees. This is aligned with my third principle of life: symbiosis (see Chapter 2). Native peoples of South America have known for generations that leafcutter ants never use the entire biomass of leaves from a single tree.[22] They always leave enough leaves so that the tree can fully recover. How is this possible? While the native peoples believe this is due to the leafcutters' respect for their surroundings (see 'The Living Theatre', page 82), a far more violent set of circumstances is at play. As soon as leafcutters begin to cut leaves, the plants activate their defence responses.[23] Specific VOCs are released that represent the plants' 'cry for help'. These VOCs are recognised by the leafcutters' predators, mostly carnivorous insects; in a sense, plants issue signals that the next meal is in sight.

But the picture is far more complex because other organisms respond to the VOCs as well, creating a complex web of ecological interrelationships. Bacteria, viruses, fungi, insects, nematodes and perhaps even birds listen in on conversations between plants and their surroundings.[24] Often, other VOCs emitted by non-plant species further enrich the conversation. This deepens the drama of the forest's evolutionary theatre – the great play of symbiosis – which started as soon as forests emerged on Earth in the Devonian geological period, 450 million years ago. The only conclusion is that millions of years of symbiotic live play, involving numerous species, resulted in stable relationships between leafcutters and trees, behind which many other species found their strategies for survival. The director who oversees all operations in the living theatre is Gaia's hyperthought. This director has exceptionally high demands: it requires that each actor produces the role of its life, and by doing so saves the species from an early evolutionary exit. That role is the unit of mind compatible with the vast Gaian mind (see Figure 3.2, page 51).

In the absence of the powerful natural director, things go astray. Take leafcutter ants again. For decades, they have been imported unintentionally from their major habitats (the forests of South and Central America) into the United States, where they often end up on farms. With no natural enemies and plenty of vegetation, the leafcutters find these new habitats to be paradise. They quickly transform from a Dr Jekyll, who in the forests of South America was constrained by bottom-up and top-down ecological forces, to an out-of-control Mr Hyde, feeding and taking over like maniacs because the ecological forces behind tropical forests have disappeared.[25] As a result, leafcutters transform from being a pivotal ecological actor that coordinates and synchronises its action within an ecological web of interdependencies, into one that acquires freedom from the ecological web and becomes a powerful pest causing big problems to human farmers and gardeners. Commercial ant killers are no match for the pest force contained within leafcutter ant colonies, and even professional pest-control measures lack an effective management formula.[26]

This example highlights the differences between a natural state of affairs and anthropogenic attempts to control nature. Some futurists argue that the distinction between the natural and the artificial is not an obstacle and that AI can faithfully simulate any natural scenario, but the complexities of the ecological living theatre and its complicated web of cross-kingdom communication challenge the power of science when it comes to modelling nature.[27] One way of addressing this problem is to consult relevant scientific disciplines outside mainstream biology. One such discipline is biosemiotics.

A Short History of Biosemiotics

Biosemiotics is the study of communication through signs – the communication we typically see in nature. Verbal communication, on the other hand, is almost exclusively the mode of communication in human culture. Biosemioticians prefer to use the designator 'language' only in the case of verbal communication, but this is shaky ground because the features of verbal communication – syntax and semantics – are also detectable in the non-verbal communication we see, for example, in bacteria and plants.[28]

According to biosemiotician Thomas Sebeok, the concept of the sign is similar to the concept of information.[29] Not surprisingly, contemporary biosemioticians frequently express gratitude to Gregory Bateson, the only biologist amongst early cyberneticists, for making a key distinction between information in the biological sense and information in the physicalist or engineering sense. Bateson's provocative description of information as the 'difference that makes a difference' fits well with the concept of the sign.[30] That is, the biosphere is made up of agents and environments that constantly exchange information; Bateson understood this better than his fellow cyberneticists. Agents require the mind to understand their surroundings. The difference that makes a difference is the summary of information in the context of the mind. As the human mind searches for meaning in any difference that makes a difference, within the confines of human cognitive space, so does the mind of the bacterial colony within bacterial cognitive space.[31]

The person who paved the way for biosemiotics, American pragmatist philosopher Charles Sanders Peirce, was in some respects similar to Bateson. While Bateson introduced the concept of meaning in cybernetics, which contributed to the argument that biology is resistant to the mechanistic explanations that work in physics (see Chapter 2), Peirce introduced the concept of meaning in semiotics.[32] Semiotics was traditionally a field of interest for linguists, and in the early days semioticians focused exclusively on the structure of the sign, ignoring meaning to the sign user. For example, the structure of the sign was considered to be dyadic: the signifier and the signified. The signifier is the vehicle of a sign; a name written on paper is a signifier for a person known by that name. The signified is the attachment of the name to the correct person in the real world. Thus, the dyadic structure of the sign, consisting of the signifier and the signified, is devoid of meaning. How do we know how to attach the right name to the right person?

Peirce introduced another element to the sign structure: the interpretant. The sign became a triadic entity. The function of the sign in the real world is not complete without interpretation of the relationship between the signifier and the signified. As the mind entered information proper in the case of Bateson's interpretation of biological information, so the mind entered the sign structure proper with Peirce's triadic semiology. Peirce also changed the terminology. The term 'signifier' was replaced by 'representamen', and 'signified' by 'object'. The triadic structure of the sign, known by the acronym ORI (object, representamen and interpretant), is the basis of biosemiotics – there can be no sign in the biological world without a three-way relationship, one that includes meaning.

Peirce's sign structure is best illustrated by an example. While mowing the lawn in my garden, I noticed several irregularities that interfered with the action of the mower. On closer inspection, these irregularities, which made the lawn's surface uneven, resembled small anthills. In triadic semiology, the object (O) is the ant colony. Its sign vehicle, or representamen (R), is the anthill. Finally, the interpretant (I) is the relationship between O and R, worked out in my mind.

How do I know that R is the signifier of the ant colony? The irregularities I observed on the surface of the lawn could also have been caused by moles. Peirce argued that R determines O by placing certain constraints on the process of signification. In our case, one of those constraints is the shape of the irregularity. The shape of an anthill is different to the shape of an irregularity caused by moles. Another constraint is the presence or absence of additional signifiers – for example, ants. I used a stick to probe the inside of the irregularity and I saw a large number of ants congregating on the spot, so the presence of ants is also an R that corresponds with O, the ant colony.

Even though I did not see the ant colony in its entirety, certain qualities of R give me confidence that my interpretation of the sign (I) is correct. I was able to gather more confirming elements in the process of signification. I used an ant-killer powder and the next day I saw dead ants on the anthill, so this is a positive relationship between O and R. Another positive relationship is the observation of a large number of winged ants on top of another irregularity on my lawn. Winged ants are part of the life cycle of the ant colony:

new queens taking off on nuptial flights. However, my interpretation of O is limited: I do not know which ant species built the anthill, although I can guess; I do not know how old the colony is, etc. Finally, the actual process by which O and R are linked in my mind, which represents the third element of the sign, I, resembles the process of translating the sign, which is a form of mental activity indicating the presence of mind in the triadic sign structure.

A person whose work made the necessary bridge between Peirce's triadic semiology and biosemiotics was a little-known German biologist who lived in what is today Estonia, Jakob von Uexküll. Uexküll was not interested in semiotics per se. Instead, he tried to interpret the biological world as if there were no humans in it – through the eyes of animals. Of course, this is not entirely possible because we are prisoners of our own minds, but Uexküll tried to cheat his mind by imagining how he would feel as a tick or a spider, and then reported this experience in a beautiful book called *A Foray into the Worlds of Animals and Humans*.[33] We have learnt from Uexküll that each species has its own perceptual world, its *Umwelt* or 'bubble', which is what I referred to earlier as the cognitive space of a species. Enter one bubble, as Uexküll tried, and you get a picture of that animal's environment and its subjective world. Move to the next bubble and the picture reconfigures itself and an entirely new subjective world appears: 'A new world arises in each bubble.'[34] The biosphere is a collection of interconnected bubbles or *Umwelten*.

The result of Uexküll's thinking was so powerful that readers such as Thomas Sebeok quickly realised the importance of the new insight.[35] It is now possible to interpret the biosphere from the perspectives of non-human agents by combining Uexküll's *Umwelt* with Peirce's triadic semiology. This peculiar combination reveals the biosphere's essential quality: extensive communication through signs. The triadic sign structure, ORI, works not only in the human *Umwelt* but also in the *Umwelten* of all organisms. Animals, plants, fungi, protists and bacteria have ways of interpreting signs that are consistent with their cognitive spaces. In this way, the biosphere becomes the semiosphere – a meaning-making process and natural epistemology that is essential to life. For this reason, my second designator for circle 1 in Figure 5.1 is the semiosphere. (At least three more fringe disciplines, including evolutionary epistemology, relational biology and systems thinking are complementary to biosemiotics. These fringe disciplines, when combined, represent a powerful challenge to the dogmas of mainstream biology.)[36]

So we can credit Uexküll for opening the route through which the mind – the process of sensing the environment and changing it through interpreting and learning – could enter the biological world. As soon as that route was opened, triadic semiology could lead to biosemiotics, because the meaning that fills the biological world is a feature of all the kingdoms of life. Indeed, Thomas Sebeok divides biosemiotics into sections corresponding to each kingdom; only zoosemiotics and phytosemiotics are relatively well established. Those dedicated to bacteria, protists and fungi, also known as protosemiotics or endosemiotics, remain underdeveloped, largely due to relatively low interest from researchers.

A Survey of Communication Modes

In the remainder of this chapter, I will present a survey of communication modes specific to each kingdom of life and describe how these modes enable interactions between kingdoms. The literature on the modes of communication in nature has exploded in the last couple of decades, and it grows by the day, so the survey I offer can be neither exhaustive nor definitive. The aim is to provide simple guiding principles so that interested readers may look for more details on their own.

Very briefly, I have classified modes of communication into physical, chemical and biological modes (see Table 5.1, page 90). Physical communication includes electrical and mechanical impulses, the senses of touch, sound and vision. Chemical communication is the use of chemical compounds such as VOCs, pheromones, hormones and specialised signalling molecules to convey messages. Finally, biological communication includes the exchange of cellular components such as pieces of DNA and RNA and subcellular structures, including viruses and plasmids in bacteria; the exchange of extracellular vesicles; and the acquisition of entire genomes in eukaryotes.

I have also included a list of kingdoms with which each kingdom communicates. This may look like an unnecessary addition because it seems likely that each kingdom communicates with every other kingdom. However, the intensity of research covering communication between different kingdoms is uneven. For example, the evidence for communication between Plantae and Fungi is now extremely strong because of the strong research focus. On the other hand, interactions between Fungi and Animalia have not been investigated with the same vigour. Those cross-kingdom interactions where the research is limited are not included in Table 5.1.

The novelty of this survey is that the reader can now look at the communication between members of the same species, or between kingdoms of life, through the prism of biosemiotics – interpreting and exchanging signs – to get an idea of how evolutionary close and distant organisms 'talk' to each other. Also, we must not forget that the web of communication is regulated from the top by the Gaian system to ensure the continuity of its autopoietic flow (see 'Gaian Science versus Human Science', page 135).

Kingdom Bacteria

To understand bacterial communication, we first need to eliminate one major misconception. Contrary to the textbook view (shaped by medical microbiology) that bacteria are disease-causing, single-cell organisms, the reality is fundamentally different. Bacteria are ecological communities – loosely organised multicell collectives – joined into a global system known variously as the bacteriosphere, the World Wide Web of genetic information or the bacterial internet.[37] This global biosystem turned the planet alive billions of years ago and has been regulating biogeochemical affairs since the

Table 5.1. Modes of communication

Kingdom of life	Mode of communication			Kingdom interactions (as covered in scientific literature)
	Physical	Chemical	Biological	
Bacteria	El, Ph	QS, VOC	HGT	Protoctista, Fungi, Plantae, Animalia
Protoctista	HW	QS	GA, FA	Bacteria, Plantae
Fungi	—	VOC	—	Bacteria, Plantae
Plantae	Touch, sound, vision, El, Ph	VOC, SM, H	GA, EV, FA	Bacteria, Fungi, Animalia
Animalia	Sound, vision, touch, hearing, El	VOC, H, P, SM	GA, EV, FA	Bacteria, plantae

Abbreviations: El - electrical impulses; EV - extracellular vesicles; FA - function acquisition; GA - genome acquisition; H - hormones; HGT - horizontal gene transfer; HW - hydrodynamic waves; P - pheromones; Ph - photosynthesis and phototaxis; QS - quorum sensing; SM - signalling molecules; VOC - volatile organic compounds.

dawn of life. As for causing diseases, the biosphere is a form of communication whereby certain microbial communities retain the habit of occasionally penetrating the borders of other organisms, irrespective of whether these are prokaryotic (through endosymbiosis) or eukaryotic (infecting plants and animals) (see biological communication, Table 5.1).

We can split bacterial communication into two categories: local and global. A typical form of local communication occurs within structures called bacterial biofilms.[38] These are bacterial social communities attached to various hard surfaces: rocks in the ocean, your teeth, kitchen sinks, etc. Within a single biofilm, all bacteria are held together by a chemical mixture secreted by the bacteria and consisting of polysaccharides, proteins, lipids and DNA. Internal biofilm structures are complex. The chemical matrix allows bacteria to grow into densely organised communities, which are separated by the network of liquid channels through which nutrients are distributed. According to the fossil record, the first bacterial biofilms emerged 3.4 billion years ago.[39] It is estimated that 40 to 80% of the global bacterial population is involved in biofilm formation.[40]

Communication within biofilms occurs through the process known as quorum sensing. When a small, free-floating bacterial community discovers a promising solid surface as a potential settlement for a larger community (i.e. a bacterial city in the form of a new biofilm), certain bacteria start transmitting chemical messages. They secrete so-called autoinducer molecules, which are biosemiotic signals that other bacteria can read and interpret. These signals are like questions addressed to all the members of the community, asking whether they are willing to embark on the adventure of building a bacterial city on this newly located hard surface. Only when the number of respondents reaches a threshold (a quorum) will the process of building a new city begin in earnest.

Throughout the city-building process, which I will explain in detail in Chapter 6, communication continues through the transmission and interpretation of chemical messages. Bacterial communities show differences in 'talking' habits. Those communities that 'like' to talk are considered extroverts, and since the greater capacity to talk leads to a better understanding of the environment, the measure of talking capacity is also the measure of bacterial IQ.[41] Communities of bacteria that talk less are dubbed introverts and have a lower bacterial IQ, relative to extrovert bacteria.

Quorum sensing is not the only form of communication within biofilms. Bacteria also communicate through electrical signals.[42] This form of communication resembles the communication between neurons in the brain. All cells, including brain cells and bacteria, possess ion channels in their membranes. Ion channels determine membranes' action potentials. Bacteria transmit electrical signals by pumping out potassium ions; the electrical signals spread through the entire biofilm. In this way, the bacterial community synchronises its metabolic habits.

Electrical signals also travel outside the biofilm. The function of long-range electrical signals is twofold. First, they reach free-swimming bacteria, which interpret incoming signals as invitations to join the biofilm community; such 'foreigners' entering an established community may bring in new genes and share them with the rest of the city dwellers through HGT. New genes bring new functions that may be useful for the city's 'infrastructure'. Second, electrical signals can reach other biofilms. In this case, the signals synchronise the feeding of two or more biofilms when a new source of nutrients becomes available. If the feeding were not synchronised, this would be detrimental for all biofilms sharing nutrients.

Given that bacteria were the first forms of life on Earth, they must have developed the capacity to read signals coming from non-living parts of the environment. A typical example is bacteria's capacity to read a portion of the electromagnetic spectrum: visible light coming from the Sun. This capacity, coupled with the metabolic capacity to use CO_2 from the atmosphere and H_2O from the surroundings, was perfected by cyanobacteria and resulted in photosynthesis – the process that has been fuelling life on Earth for the last 3.4 billion years.

As soon as biofilms started communicating with each other, a web of interconnected biofilms sprung up all over the planet. This web, growing in oceanic, terrestrial and underground habitats, combined with other bacterial communities, such as free-floating bacteria in the atmosphere, after a certain period of time (a few hundred million years or so) turned into the global bacterial community, or the bacteriosphere. This was the first form of Gaia, the planetary microbial system that could regulate itself.

In the epic process of overwhelming the planet, the bacterial propensity to swap DNA was essential. Thus, the web of bacterial communication was enriched by a new form of communication: biological communication. A typical example of biological communication is horizontal gene transfer, or HGT. Bacteria can pick up any piece of DNA from the environment and integrate it into its genome. The integrated piece of genetic information can be transferred to other bacteria, becoming part of the bacterial pan-genome. A far-sighted microbiologist, Sorin Sonea, named the totality of bacterial genes the World Wide Web of genetic information – a truly global market on which bacteria trade genes.[43]

Bacteria also exchange genetic information via viruses and plasmids. Viruses are the most numerous biogenic structures on Earth.[44] The virosphere – the community of all viruses on the planet – is dominated by bacterial viruses called bacteriophages, or simply phages. The bacteriosphere and the virosphere are structurally coupled and represent the greatest genetic mosaic on Earth. (Interestingly, even viruses – biogenic forms incapable of living on their own – communicate with each other through the chemical arbitrium.[45]) Plasmids, on the other hand, are circular pieces of DNA present in multiple copies in many bacteria, and they swap frequently between them. Plasmids are different from the main bacterial genome, a piece of DNA called a genophore.[46]

The function of the bacterial global genetic market was to transmit biological function. Bacteria that share DNA have the same set of proteins, which in turn facilitate biological functions such as nutrient cycling, chemical communication, phototaxis or biofilm construction. In this way, a variety of biological functions spreads globally; bacterial communities exchange genes as carriers of specific functions. Transmitting biological function was the global process regulated from the top, by the Gaian system, that integrated all local bacterial communities into the bacteriosphere.

Analysis of the bacterial genetic market conducted by Sorin Sonea, Carl Woese and their colleagues showed that the bacteriosphere is a biological system dominated by the Lamarckian rather than the Darwinian paradigm (see epigraph, page 79).[47] Darwinian principles operate only within communities of reproductively isolated organisms, such as plants and animals, that have complex and stable genomes transmitted vertically from parents to offspring. Bacteria, on the other hand, are not reproductively isolated; they lack stable genomes and share DNA extensively through HGT. Sonea and Woese argued that bacteria do not fulfil species criteria.[48] Bacterial classification into species is a useful human artefact, rather than the reflection of biological reality.

Like Sonea and Woese, evolutionary biologists Eugene V. Koonin and Yuri I. Wolf argued that HGT is a process that follows Lamarckian rather than Darwinian principles.[49] The first Lamarckian principle, according to Koonin and Wolf, is that environmental factors may induce genetic changes that become heritable. The second principle is that these changes might not be random because they target specific genes. The third principle is that the induced changes will provide adaptation to the original causative factor from the environment. None of these principles are compatible with Darwinism, which holds that environmental factors cannot direct genetic changes and that genetic changes are entirely random and fixed by natural selection.

Here is an example of an HGT process that operates along Lamarckian lines.[50] It is well known that bacteria frequently acquire resistance to antibiotics, a situation that poses major medical problems. Most bacteria exposed to a given antibiotic will die. However, rare bacteria will acquire genes for antibiotic resistance, usually via plasmids from other bacteria, via HGT. The surviving bacteria will produce colonies of antibiotic-resistant bacteria that will reinforce the resistance. Thus, the causative chain of events induced by the environment – a bacterial response to a problem of antibiotic toxicity, administered by the practices of medical microbiology – leads to the solution of the same problem.

Given that HGT is a dominant mode of genetic exchange in bacteria – the only life forms on Earth for the first two billion years of evolution – Lamarckism as a mode of evolution must have been dominant throughout this period. The trouble is that biologists remain resistant to Lamarckism. There are multiple reasons for this, from Lamarck's incorrect interpretation of the inheritance of acquired characteristics (such as the infamous story of giraffes and long necks), to the misuse of Lamarckism by the likes of Paul Kammerer and Trofim Lysenko. However, the picture is changing. New research suggests that the Lamarckian paradigm is legitimate in certain cases. It seems likely that in the biology of the future, Darwinian and Lamarckian paradigms will merge.[51]

Kingdom Protoctista

Single-cell eukaryotic microbes, usually referred to as protists, emerged roughly 1.5 billion years ago when certain prokaryotic microbes learnt how to live inside one another through endosymbiosis.[52] Their systematics and the systematics of their descendants are still a matter of debate. The result is a distinct kingdom of life, Protoctista, which refers to these unicellular eukaryotic microbes and their descendants, some of which are multicellular. Protoctista are classified into thirty phyla, ranging from algae and seaweed to slime moulds, slime nets and protozoa, such as paramecium. The term 'protist' traditionally refers specifically to eukaryotic unicellular microbes.

The emergence of new life forms changed the game of communication profoundly. Bacteria, the only inhabitants of Earth for two billion years, got their first alien neighbours, protists, with whom they shared the pleasures and troubles of neighbourly relationships. This also meant that cross-kingdom communication appeared for the first time in the history of life. The practical consequence of these changes was the explosion of symbiotic relationships beyond endosymbiosis: mutualism, commensalism, parasitism, predation and competition.[53]

The newly emerging symbioses required specific forms of communication between hosts and symbionts in order to survive the filter of natural selection. A typical example of symbiosis between bacteria and protists is predatory symbiosis. Many protists are natural predators of bacteria. The process of predation depends on bacterial VOCs.[54] These chemical messages inform protists that grazing fields of bacteria (i.e. biofilms) are in the vicinity. Protists are expert readers of bacterial VOCs. They can distinguish between different bacteria and graze them preferentially. However, bacteria quickly developed resistance to grazing. Under the influence of alarm signals transmitted through quorum sensing, bacterial cities can quickly switch into combative mode to fight protist predators.

In the particular relationships between bacterial and protist *Umwelten*, new and previously unknown scenarios arose. For example, *Paulinella* is a genus consisting of several ameboid species.[55] Some *Paulinella* species merged with cyanobacteria 90 to 140 million years ago, in an event that made *Paulinella* capable of photosynthesis. This surprised the scientists who discovered it because it occurred more than a billion years after the original endosymbiosis that resulted in the emergence of the first eukaryotes. In the subsequent endosymbiosis, the host and symbiont exchanged genes through HGT. So here we have examples of two forms of biological communication: *Paulinella* acquired the entire genomes of cyanobacteria, and then these genomes were remodelled through HGT.

A range of other symbioses between protists and bacteria have been identified, many of which involve biological communication. One of the most spectacular examples is the farming symbiosis between the social amoeba *Dyctiostelium discoideum* and bacteria from the genus *Burkholderia* (see Chapter 10). In brief, the amoebas act as 'farmers', who grow bacteria in tiny gardens and harvest them as food.[56] Another example of biological communication is the sexual reproduction of the protists *Salpingoeca rosetta*, facilitated by the bacterium *Vibrio fischeri*.[57] This bacterium secretes the 'aphrodisiac' protein aptly called EroS, which acts as the mating signal for *Salpingoeca*.

Further examples of biological communication include locomotion. The movement of some protists is helped by bacterial symbionts, which greatly enhance the mobility of their hosts. For example, the entire body of the *Mixotricha paradoxa*, a protozoan that lives in termite hindguts, is covered by a collective motor of 200,000 spirochaete bacteria.[58] *Mixotricha*'s own four flagella serve not as a locomotion device but as a steering device. In *Mixotricha*'s *Umwelt*, perhaps, spirochaete power is the equivalent of horsepower. *Mixotricha*, also known as 'the protist beast', has no fewer than five genomes. Lynn Margulis and Dorion Sagan thought that *Mixotricha* should be the poster animal for symbiogenesis.[59]

Locomotion assistance sometimes involves electromagnetic phenomena. For example, magnetotactic bacteria, symbionts of some euglenoid protists, guide their hosts through magnetic fields and lead them to rich meals of anoxic marine sediments.[60] There are many other examples of protist–bacteria symbioses, including metabolic, defensive and parasitic types, each one requiring a specific mode of communication between hosts and symbionts.

When protists are examined on their own, outside the context of symbiotic relationships with bacteria, they show typical features of the sociality we have seen in bacteria. Many ameboid species are social organisms that rely on quorum sensing to communicate with each other.[61] For example, when food is scarce, *Dictyostelium discoideum* (aka slime mould) switches from a vegetative state, in which the collective of unicellular amoebas feeds on bacteria, to an aggregation state, in which unicellular amoebas aggregate into a slug, from which a fruiting body emerges to disperse spores. Quorum sensing, as a means of communication, is used in both vegetative and aggregation states in the life cycle of such social amoebas.

Scientists from Stanford University have recently discovered a unique mode of communication in the protist *Spirostomum ambiguum*.[62] *Spirostomum* is a single-cell organism, visible to the naked eye, which has the amazing ability to contract its body very quickly in the presence of a predator. Within a few milliseconds, *Spirostomum* can contract its 4 mm body length by two-thirds. This rapid contraction generates vortex flows that neighbouring cells read as messages also to contract, leading to a wave of contraction within the *Spirostomum* community (along with the simultaneous release of predator-detracting toxins). When scientists observed this unusual protist dance triggered by hydrodynamic waves, one of them was so fascinated that he commented: 'we really don't know what life is capable of'.[63]

Protists use other forms of physical communication, too. The slime mould *Physarum polycephalum* is a large single cell. It forms long protrusions, a network of protoplasmic tubes, that move in the direction of a food source. The tube movement is regulated by electrical impulses.[64] It is not known whether the electrical impulses used *within* the long unicellular body of *Physarum polycephalum* are also used in communication *between* individual protists. Some scientists speculate that the spectrum of communication modes in protists is far greater than the current research can show.

Another peculiarity of protists is that they are transitional forms between single-cell organisms and multicell organisms. We have seen that social amoebas can switch from single-cell existence into multicell forms. This is driven by quorum sensing, as well as bacterial VOCs.[65] When bacteria, on which amoebas feed, are plentiful, amoebas remain unicellular organisms. When bacterial grazing fields are depleted, however, unicellular amoebas merge into a multicellular fruiting body that disperses into dormant spores, only to become activated again as unicellular amoebas when the concentration of bacteria rises. This is a typical example of biosemiotics at work within the social amoeba's *Umwelt* through sensing and interpreting environmental cues and planning for the future.

The importance of communication in the switch to multicellularity – the emergence of multicellular fungi, plants and animals – is illustrated by the finding that *Monosiga brevicollis*, one of the closest living protist relatives of animals, possess more signalling molecules than any known organism.[66] These molecules, called tyrosine kinases, are biosemiotic signs for intercellular communication, and are produced by protists as well as animal cells. This provides an important clue for understanding the evolutionary transition from unicellular life to its multicellular equivalent.

Kingdoms Fungi, Plantae and Animalia

Scientists are not clear on when exactly multicellularity first emerged, because of discrepancies between the fossil record and the dating of evolutionary events using molecular clocks. However, the growing consensus is that multicellularity in the form of multicellular fungi, plants and animals emerged less than one billion years ago. Protists such as *Monosiga brevicollis* and *Salpingoeca rosetta* invented multicellularity by playing with signalling molecules.[67] This major evolutionary event, one of the eight major evolutionary transitions, is the only one that can be replicated in the laboratory.[68] Scientists have been able to turn yeast, a unicellular fungus, into a multicellular body by selecting certain big cells, which then started merging on their own.[69]

The new multicellular life forms immediately integrated into the existing ecological web, in which new symbioses between unicellular and multicellular life forms exploded. We have only started discovering the richness of these interactions. Each one adds a new strand to the web of communication that covers Gaia inside and out, which I like to refer to as 'physiology of mind' (see Figure 2.1, page 44).

The transition to multicellularity included an adjustment of communication modes. While the sociality in bacteria and protists required forms of communication that easily jumped from one group of cells to another, with no barriers imposed by differences in cell anatomy, the emergence of multicellular bodies changed the nature of signal transduction. Cells forming the multicellular bodies of fungi, plants and animals sacrificed individuality for corporate individuality.[70] Communication within corporate bodies – genetic clones of cells descending from a single chimera produced by sex – was restricted to signal transduction between cellular clones, or exchange of extracellular vesicles. But genetic clones were not anatomical clones. A new type of genetics, called epigenetics, established how cells control gene activity without changing the DNA sequence, thus allowing for the emergence of a few hundred cell types in plant and animal bodies. In animals, these range from long, thin brain cells that cannot move, to the round, mobile cells of the immune system, and in plants from the tracheid cells that enable water transport through xylem, to the sclerenchyma cells that make plant bodies hard.

Corporate bodies imposed strict control on the behaviour of trillions of cells under their command. Dissenters are punished by death. The execution process is called apoptosis or programmed cell death. It is communicated by a set of well-defined chemical messages transmitted by a group of proteins called caspases.[71] Invasion of animal corporate bodies by foreign cells, such as bacteria, is fended off by crews of mobile cells. These are patrols that travel throughout the body to identify invaders and kill them on the spot. If the invaders are more numerous or more powerful, the patrol crews send back chemical messages in order to mobilise the entire cellular army of lymphocytes.

In summary, intercellular communication within corporate bodies involves electrical impulses (e.g. brain cells), chemical signals (e.g. tyrosine kinases, hormones) and biological communication (e.g. exchange of extracellular vesicles).

Corporate bodies also acquired the senses of vision, touch, smell, taste and hearing. In animals, the senses are centrally coordinated by the brain. In plants, the coordination of senses is less centralised. The senses differ dramatically between plants and animals. Animals' eyes are precision instruments that enable visual representation of the world. Plants, on the other hand, have ocelli on their leaves; these are less complex vision instruments that help plants find space with respect to neighbours.[72]

The *Umwelten* of some ocean-dwelling animals are light-dominated worlds. For example, corals, some sea turtles and many fish species communicate using biofluorescence.[73] Their bodies absorb electromagnetic radiation at one wavelength and emit it at a different wavelength, resulting in spectacular green, red and orange patterns. A similar form of biofluorescent communication is used by butterflies, parrots, spiders and even flowers.[74] Sound communication is widespread in termites, birds, dolphins and whales. Scientists are so impressed with the language of sperm whales, coded in sound patterns and consisting of clicks transmitted in rhythmic series called codas, that a project called Project CETI (the Cetacean Translation Initiative) was launched to decode the language of sperm whales using AI, robotics, linguistics and big-data analysis.[75] Plants also use bioacoustics to coordinate pollen release with insects in a process called buzz pollination.[76]

But the freedom of corporate bodies within the Gaian system was restricted from day one. Fungi, plants and animals emerged from the ecological web formed by bacteria and protists, and they remain entangled in that web. There are no plants or animals without accompanying microbiota, which include bacteria, protists and fungi.[77] Corporate bodies and their microbiota are referred to as holobionts, but holobionts are the tip of the iceberg of ecological interdependencies. Earlier in this chapter, I used the expression 'the great play of symbiosis' to describe the web of ecological interdependencies in the context of the relationship between trees and leaf-cutter ants. This great play – a living theatre being staged before our eyes and growing richer by the day with surprising discoveries that challenge anthropocentric dogma – surpasses the great Shakespearean dramas, for Shakespeare overestimated the originality of the human *Umwelt*.

I've only scratched the surface of the large body of communication amongst the kingdoms of life within the body of Gaia, not to mention undiscovered modes of communication, some of which will probably remain forever beyond the grasp of science.

Who would think that in only the past few decades scientists have discovered a new frontier in the underground network of fungi and bacteria that connects aboveground forests in an incredible web of interconnection? Dubbed the Wood Wide Web by scientists, this massive and powerful network preceded the internet by hundreds of millions of years.[78] In a typical mutualistic symbiosis, trees produce bountiful sugars through photosynthesis and make some of these sugars available to fungi. In return, the fungi colonise the trees' roots systems and provide them with essential minerals that the fungi extract from the soil. The power of this symbiosis goes beyond mutualistic nutrition. It also results in a powerful underground network of interconnected fungi that plants then use to exchange messages with other plants.

The Wood Wide Web, which likely emerged around the time forests emerged, about 400 million years ago, coincided with the development of a rich *Umwelt* of trees. This *Umwelt* shares many characteristics of human culture. For example, big trees take care of small trees, like parents taking care of their children, by directing nutrients through the Wood Wide Web. Indeed, there is widespread kin recognition in plants.[79] There are examples of great acts of altruism in this *Umwelt*, such as old trees 'informing' younger relatives that they will die soon so that youngsters can 'inherit' their nutrients. But there are also games of deception and even wars in this *Umwelt*, facilitated by the Wood Wide Web messaging platform.

Naturally, there are also abundant examples of cross-kingdom communication between animals and other kingdoms of life, such as the leafcutter ants in South American forests: these ants practise a version of agriculture in which they farm fungi through biological communication, not unlike how humans farm certain cereal plants (see Chapter 10). New research shows that ants do not behave like automata; they show signs of individuality. This means that communication with other ants, and possibly other species, could be affected by their character.

Take, for another example, a situation involving animals, protists and bacteria.[80] Termites are insects that form large societies, called superorganisms, which are housed in self-built structures called termite mounds.

One Australian termite species, *Mastotermes darwiniensis*, houses the protist *Mixotricha paradoxa* in its gut. The relationship between the termite and the protist is mutualistic. The protist helps the termite digest the cellulose and lignin in its wood-based diet, and the inside of the termite gut is a safe environment for the protist. *Mixotricha* also houses a large team of bacteria on its body to aid its locomotion (ectosymbiosis), and at least one bacterial species inside its body (endosymbiosis), which digests cellulose. So the nested drama hidden behind the walls of termite mounds has a total of four biological levels connected through various forms of communication: the termite superorganism, individual termites, the protists in the guts of termites and the bacteria living inside and outside the protists' bodies. They are all connected by the need of the termite superorganism to feed itself. Gaia is a multitude of nested worlds.

The Talking Planet

The short survey of communication modes summarised in Table 5.1 (page 90) shows that the planet is reverberating with chats, laughs, cries, shouts, light flashes, buzzes, bursts of electricity, water waves and sounds of bodies and body parts merging, produced by its numerous citizens, including bacteria, protists, fungi, plants and animals. There are millions of languages spoken in this huge living theatre. Its stage covers all ocean and terrestrial habitats, and it goes 70 kilometres into the air and dozens of kilometres underground.

We humans, the youngest actors in the theatre, integrate the experiences of many other actors, some of whom make up our bodies. These experiences open our eyes, ears and minds, and allow us to go back to the origin of the theatre and watch its evolution retrospectively. Despite our skills and powers, we remain supporting actors. How far we are able to reach into the future, and whether we will turn into a more important actor, depends on our abilities to understand subtle messages from the powerful director in charge of the play: Gaia's hyperthought (see Figure 2.1, page 44).

CHAPTER SIX

Engineers

Whoever wants to hold on to the conviction that all living things are only machines should abandon all hope of glimpsing their environments.

JAKOB VON UEXKÜLL[1]

Engineering is as common as communication, and the planet is a permanent yet ever-changing building site. Everything started with the invisible engineers: bacteria. They have built the stage of a living theatre so all organisms from the remaining four kingdoms of life can perform their roles in the endless play of evolution. Today we are witnessing our struggle to understand the play. Everything in it is ephemeral; biocivilisations come and go. The stage remains the only permanent structure. Even planetwide catastrophes (we know of six so-called major evolutionary extinctions) couldn't damage the stage irreversibly.[2]

The play itself is full of feats of engineering: houses built by protists; metropolises built by ants and termites; the planetary communication network built by bacteria; tools invented by animals; the precision of surveying flower fields by honeybees; dead bacterial metropolises frozen in time like Pompeii; the plant-dominated engineering sites called biomes; medical devices invented by bacteria to protect them from viruses; large biogenic constructions ranging from fairy circles in Namibia and Mima mounds in North America to *murundus* in Brazil and *heuweltjies* in South Africa, built by the joint efforts of plant and insect biocivilisations; ecological engineering performed by fungi, insects and snails; and many, many more.[3] Planetwide natural engineering has never stopped throughout the entire existence of the Gaian system.

Humans follow, rather than set, this pattern. Engineering is so important for our civilisation that we have reserved a hallowed place for it in modern education. In the acronym STEM (science, technology, engineering and medicine), used by universities worldwide to attract students, the third letter is a testament to the assertion that engineering is essential for the progress of human culture. Forms of engineering are so numerous that it is difficult to produce a definitive list, but they include mechanical, electrical, aerospace, nuclear, genetic, environmental, natural, chemical, computer, ecological, sustainable, green, geotechnical, biological, river, civil, energy, life cycle, climate and biomedical engineering.[4] The engineering feats of other species have become a good source of human innovation, and this copying of natural engineering for the benefit of human culture is called biomimetics. It was established in the mid-20th century by biophysicists such as Otto Schmitt, and perfected by his followers.[5]

There is a point of contention about the differences between human and natural engineering. Human engineering is fully computable; before we build anything, we produce mathematically precise blueprints. Blueprints are not required, however, in natural engineering. While there are elements of mathematical precision in some forms of natural engineering, like the famous honeybee algorithm, natural engineering as a whole is not based on pure computation. Life is more than a machine and may resist human attempts to fabricate it.[6] But many scientists and engineers are adamant that we can. Not only that: science-based predictions suggest that we humans can alter the planetary biosphere, the solar system, even the galaxy and beyond.[7]

The future will show who is right. Or maybe not. The time required to test human engineering may be incommensurable with our ephemeral evolutionary status. Yet we can place human engineering in the context of biocivilisations. This may help us understand the limits of our engineering skills relative to the rest of life.

What Is Engineering?

Even though engineering was used by early humans – from inventing primitive tools to constructing prehistoric monuments – contemporary interpretation of engineering is heavily biased towards modern inventions. If you visit engineering departments at universities worldwide, professors will tell you that engineering is any form of precision construction that relies on science.[8] The origins of modern science can be traced to 17th-century Western Europe. This means that our modern understanding of engineering is biased towards the last 300 to 400 years of Western civilisation. While this narrow window of time might not be a problem for anthropology, such narrow investigative horizons pose a significant problem when we try to place engineering in the context of nature. How can we reconcile modern human engineering practices, from nuclear to civil engineering, with the practices of ecological engineering performed by insects, or natural genetic engineering performed by bacteria?[9]

To place engineering in the context of biocivilisations, we need to break off these anthropocentric shackles that narrow our investigative horizons. We have to identify a second designator for engineering. That designator must be wide enough to incorporate all components of engineering as a human activity, but universal enough to break the anthropocentric walls that make us blind to the work of other species. We have already done this in Chapter 5 in naming the second designator for communication the semiosphere, by identifying biosemiotics as a discipline that universalises the concept of biological communication (see Chapter 5 and Figure 5.1, page 81). Using a similar approach, I will now show how technology can become the second designator for engineering.

Philosophers of technology have identified a problem with defining technology. The concept of technology is so wide that when practitioners in different fields and with different understandings put their stamps on it, it becomes difficult to share with other practitioners. For example, the meaning of technology is different for information scientists and lawyers, or for high-tech industries such as robotics and farming in the milk industry.

This problem of defining technology was solved elegantly by philosopher Richard Li-Hua. He collated as much detail as possible from different branches of technology and unified all their components into a single system, universal enough to reflect all forms of technology. The result was so impressive that we can now use it for our purpose: to incorporate human engineering, but also to extend engineering into the realm of biocivilisations. However, the relationship between engineering and technology is only a starting point in investigating the meaning of engineering in the context of biocivilisations. We will need to introduce certain philosophical interpretations of technology to make the picture more complete. Before we go there, let's examine Li-Hua's definition of technology.

In Li-Hua's vision, technology has four components: technique, knowledge, organisation of production and product (see Figure 6.1).[10] The technique consists of instruments (tools and machines), materials and the

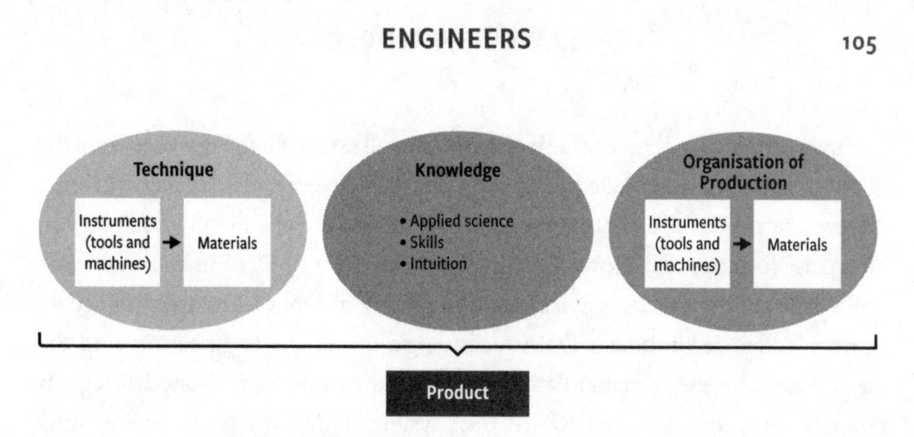

Figure 6.1. Richard Li-Hua's vision of technology.

method for bringing those instruments and materials together. Knowledge consists of applied science, skills and intuition; Li-Hua sees knowledge as the key component of technology. The organisation of production is the idea that technique and knowledge must be organised before they can bring about effective results. The product integrates technique, knowledge and the organisation of production.

We can now use examples from human engineering to test whether the concept of technology, as defined by Li-Hua, can incorporate all components of engineering and thus serve as its wider basis. Let's examine the area of biology known as genomics, which relies heavily on genetic engineering. Genomics, as an academic discipline, has been thriving since the inception of the Human Genome Project (HGP) in the early 1990s, which had the goal of sequencing the entire human genome.[11] The material for genomics is the genome of any organism we select to analyse and manipulate. DNA sequencing means reading precisely four letters of the DNA alphabet – adenine, cytosine, guanine and thymine – in the string of three billion letters that comprise the human genome. To achieve this, scientists cut human DNA into millions of small fragments and used genetic engineering to place these fragments into viral, bacterial or yeast vectors to produce a library of human DNA fragments representing the entire genome.[12] Each fragment was then sequenced and placed on the map of the human genome. The expertise scientists gained during the HGP allowed them to start sequencing the genomes of other organisms as well. Since then, thousands of animal and plant genomes have been sequenced, and bioscientists are also trying to sequence the genomes of all eukaryotes in a project named the Earth BioGenome Project.[13] There are also parallel efforts to sequence one million human genomes for medical purposes.[14]

Instruments for genetic manipulation are well established. Fifty years ago, the first 'genetic scissors', called restriction enzymes, were isolated from bacteria. These are precision tools that cut DNA from one organism – say, bacteria – at a precise location and allow inserting the same piece of DNA into the genome of a different organism – say, a plant. The mixing of DNA from different organisms is called recombinant DNA technology, or genetic engineering, and the tools for such genetic manipulation are becoming more sophisticated. Scientists have recently invented the CRISPR-Cas9 system, a powerful gene-manipulating tool that can target virtually any gene in any genome and modify or replace it with a high degree of precision. The predictions are that CRISPR-Cas9 will eventually be applied to treating genetic diseases in humans.

The tool for the precise reading of individual DNA letters is called DNA sequencing. The technique was invented in the 1970s and was used heavily during the HGP under the name Sanger sequencing.[15] After the completion of the HGP, a new sequencing technique was invented, named Whole Genome Sequencing or Next Generation Sequencing.[16] It is much faster and cheaper than the old-fashioned Sanger sequencing because it relies on automated procedures and big computing power. In the early 1990s, scientists realised that they could bring together instruments (restriction enzymes and vectors) and material (any given genome) for the HGP, resulting in the birth of the technique.

The knowledge of genomics is the accumulated experience in the analysis of nucleic acids, best described by the phrases 'molecular biology' and 'molecular genetics'.[17] Key components of this knowledge are: (1) understanding the structure and function of DNA, (2) understanding the mechanisms of DNA organisation within cells and how the information stored in DNA is used for the synthesis of proteins and (3) understanding the methodology behind DNA isolation from cells, and the subsequent manipulation and analysis of DNA to identify the genetic code in the form of long strings of DNA letters that form the genomes of every organism.

The organisation of production, in the case of the HGP, was similar to any other industry. As soon as the money for the project was allocated to scientists by the governments of industrialised nations, an international research consortium was established. Initial work consisted of producing genetic and physical maps of the human genome and 'tracing the territory' by placing appropriate landmarks in the form of genetic markers. When the landmarks became dense enough, the territorial gaps between them were filled by an industrial process of decoding DNA sequences performed by robots (automated Sanger sequencing). According to Craig Venter, one of the participants in the HGP, the industrial process was too slow. He had an idea for how to speed it up. Private investors liked the idea and the project split into two: public and private. The ensuing competition resulted in the earlier completion of the HGP.[18]

The expected product was a complete sequence of the human genome of approximately three billion letters using the four-letter alphabet. However, small parts of the genome could not be precisely sequenced, so the final product was an incomplete but acceptable sequence of the human genome, divided into twenty-four volumes. Each volume represents an individual human chromosome. The longest volume is chromosome 1, which consists of about 250 million letters. The shortest volume is chromosome 22, which consists of about 50 million letters. The meaning of the entire 'book' of the human genome remains unknown. Of the total 20,000 to 25,000 genes (protein-coding sequences) in the human genome, the identity and function of many remain unknown. Nor is it known how the genome operates in about 200 types of cells in the human body.

The example of the HGP as a project that can be placed into the realm of genetic or biological engineering shows that it fits Li-Hua's concept of technology. We can apply Li-Hua's principles to understand any engineering project, from building the ancient Egyptian pyramids to designing the Clifton Suspension Bridge (the first suspended bridge, designed by the great British engineer Isambard Kingdom Brunel in the 19th century). We can do the same thing for massive applied-science projects such as the Manhattan Project, ENCODE and the Earth BioGenome Project. In all cases, you can map engineering components onto the platform that represents technology.[19]

There is one important difference, however, between ancient and modern engineering. The Egyptian pyramids were constructed 4,000 years before the invention of modern science. This suggests that science (at least as we know it) is not required for engineering. Yet engineers who analysed the structure of the Great Pyramid of Giza were impressed by the precision of the measurements and the quality of the construction in the absence of the tools and machinery available to modern engineering companies.

How could this be? In Li-Hua's definition, technology includes types of knowledge, such as intuition and skills, that are different than what we classify as applied science today. It is through this open door that we can also examine engineering in the context of biocivilisations. As we know from the second principle of life (agency), every organism is an agent. Agents know their environments and constantly change them using sensory and effector organs.

Social Insects

The key feature of human engineering is planning, or the creation of blueprints before the construction process starts. As far as we understand animal cognition, sophisticated, human-like planning is absent, or at best exceptionally rare. We can safely say that there are no blueprints for constructions in cases of animal engineering. Yet animals can produce extremely complex constructions. For example, social insects follow simple biosemiotic rules that allow them to form societies (decentralised systems capable of changing environments by reading and interpreting simple environmental cues).[20]

Ethologists have identified four types of behaviour amongst social insects that allow them to construct nests of amazing complexity, yet none of these behaviours include planning. The nests range from termite mounds several metres high, with sophisticated interiors including the control of air circulation, to underground nests built by leafcutter ants that contain networks of internal transportation tunnels connecting factories for food production.[21] The four types of social insect behaviour that facilitate complex constructions include: (1) reliance on templates, (2) stigmergy, (3) self-organisation and (4) self-assembly.[22]

Reliance on templates means that the 'blueprints' for nests already exist in the environment in the form of environmental heterogeneities. Insects only need to follow simple environmental cues such as temperature and humidity gradients to build nests and spatially distribute eggs, larvae and pupae. Insects also rely on templates produced by other insects. For example, queens emit pheromones that diffuse and create templates in the form of decreasing gradients. Termite workers follow the pheromonal gradients to distribute soil pellets. In this way, the shape of the nest gradually emerges.

Stigmergy is a phenomenon described by French zoologist Pierre-Paul Grassé that allows for the coordination of work.[23] Typically, an activity performed by a worker insect leaves traces that stimulate other workers to perform similar acts. Stigmergy ranges from stimulation through pheromone trails to changing the pattern of soil pellet configuration in the nest structure. The phenomenon of stigmergy shows that building activities are not guided by individual workers but instead emerge from social interactions.

Self-organisation means that the system-level pattern (the nest structure) emerges from the interactions of the lower-level components (the worker insects), which rely only on local information, not on the global pattern. Self-organisation is governed by positive and negative feedback loops, amplification of certain behavioural patterns and multiple agent interactions.

Finally, self-assembly, also known as qualitative stigmergy, is the response of social insects to certain qualitative stimuli. While self-organisation is a response to quantitative stimuli, such as temperature, humidity or pheromone gradients, self-assembly allows individual insects to solve construction problems through qualitative signals. For example, when social wasps are faced with the problem of where to add a new cell, the problem is solved more easily when wasps encounter the three-wall pattern in the growing nest structure, rather than the two-wall pattern. The hexagonal cell fits more easily into the three-wall structure of existing hexagons than into the two-wall structure (see Figure 6.2).

We can now go back to Li-Hua's technological platform (see Figure 6.1) and map the processes of nest construction by social insects onto it. The bodies of social insects act as sensitive instruments (tools and machinery) equipped with chemoreceptors, mechanoreceptors, thermoreceptors and hygroreceptors.[24] These versatile body instruments search for suitable materials available in the environment, such as soil and water. For example, ants and termites carry soil pellets in their mandibles and deposit them at nest

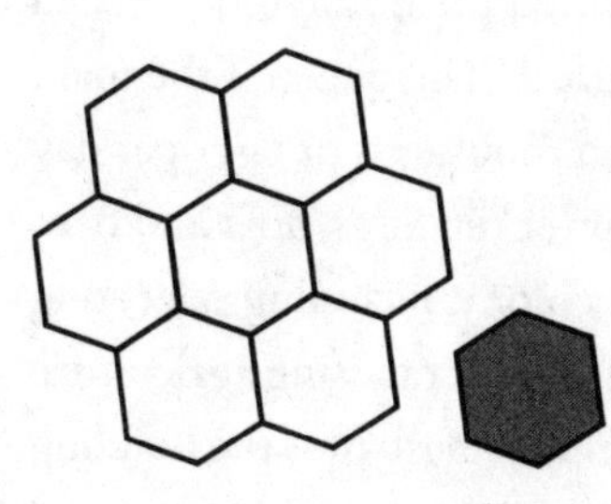

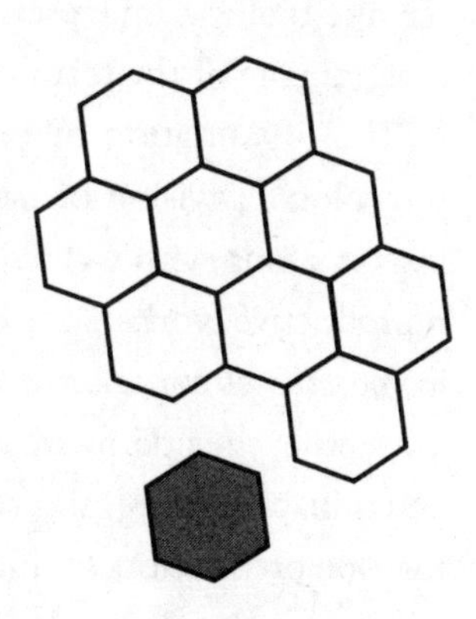

Figure 6.2. New cells (grey) facing either two-wall or three-wall structures.

sites, and water is important for maintaining specific degrees of humidity within termite nests. Entomologists noticed mass 'kissing' in termite colonies; this process of distributing water mouth-to-mouth serves to regulate humidity within the nest.[25] Other materials include pieces of chewed-up wood that social wasps stick together to make hexagonal cells, the building blocks of their nests.[26] Honeybees, on the other hand, have glands that secrete wax, from which the honeycomb – a mass of hexagonal cells that comprise honeybee nests – is created.[27] Thus the technique emerges spontaneously: insects' bodies, or instruments, follow simple environmental cues in the search for suitable materials.

Knowledge, in the case of social insects, can be classified as a form of intuition. It includes all those processes that contribute to nest construction, including the reliance on templates, stigmergy, self-organisation and self-assembly. Entomologists frequently characterise social insect colonies as superorganisms.[28] Indeed, cognitive aspects of superorganisms can be characterised as the collective minds of colonies.

Manifestations of the collective mind include not only the system-level coordination of processes behind nest building, but also the processes behind nest maintenance after the building is complete. For example, feeding the insect colony through agricultural practices (see Chapter 10) represents the superorganism's digestive tract, and the bubble-like chambers and branching air passages that occupy large parts of termite mounds are analogous to the superorganism's lungs. These chambers constantly inhale and exhale gases and thus ensure a stable nest environment. J. Scott Turner, an expert in termite biology, interprets termite mounds as 'living' structures that are an integral part of the termite colony.[29]

The organisation of production stems from the social organisation of the colony. Division of labour is a well-documented phenomenon in social insects, along with well-characterised caste systems.[30] The queen is the main reproductive worker, producing hundreds or even thousands of eggs per day in the case of some termite and ant species. Worker termites, ants and bees are sterile; they perform all the tasks within the colony, including collecting and transporting eggs, taking care of larvae and pupae, practising agriculture for food production, building the nest, maintaining the nest after the building is completed and carrying out all other tasks required for the colony to function as a superorganism.

The product is a nest that behaves as a dynamic system. It takes four to five years for termites to build a nest, and it lasts for as long as the queen is alive, which varies from fifteen to thirty years (whereas worker termites live only one to two years). Life expectancies of ant queens and ant colonies are similar to those of termites.[31] Honeybee queens, on the other hand, live only one to two years.[32] Maintaining nests – in particular termite and ant nests, which may last for several decades – is an enormous task. Workers and soldiers are faced with challenges ranging from weather (e.g. heavy rainfall can greatly damage a nest) to predators. Everyday housekeeping duties include regulating humidity and air circulation.

Social insects are not the only practitioners of technological behaviour amongst animals. For example, birds invent tools, as well as full-blown technologies recognised by four components from Figure 6.1.[33] Likewise, plants, protists, fungi and bacteria all have their technologies (see 'How Bacteria Construct Cities', page 116, and 'Construction of Forest Infrastructure', page 118). We can conclude that the concept of technology, as defined by Li-Hua, (1) captures all elements of human engineering practices and (2) allows us to identify non-human engineering practices. The next task is to focus on the meaning of technology in the context of the Gaian system. This will allow us to integrate human technologies with technologies present in all kingdoms of life. Here, we need help from philosophy.

The Question Concerning Technology

In 1954, the controversial German philosopher Martin Heidegger (controversial mostly due to his well-documented links to Nazism) published 'The Question Concerning Technology', an essay that represents one of the most important challenges ever issued on the meaning of technology.[34] Heidegger's key argument was that the idea of control over man-made technological processes is a delusion. The belief that industrialised technological projects could ever allow us to control planetary genetic resources, even for so-called beneficial reasons, is problematic.[35] The failure to understand the essence of technology may result in the loss of what makes us human; it is equally likely to transform human beings into technological slaves, unable to control our own lives. The Gaian river may swallow our instruments, machines and technologies and sink them to its riverbed. These rusty remnants will become archaeological deposits, signs of the world being misunderstood. Yet Heidegger's vision of technology is not pessimistic. The dangers associated with modern technologies also stimulate the power and possibility of changing course, or what Heidegger referred to as 'saving power'.

Before we focus on Heidegger's interpretation of technology, it's important to contextualise it within Li-Hua's framework and the broader concept of biocivilisations itself. Li-Hua's framework doesn't tell us anything about the dangers of modern technologies, but within it we can use the third component of technology – knowledge – to incorporate the factors that Heidegger considered important: the essence of technology and the dangers arising from modern technologies in particular. The way to achieve this is to add another form of knowledge to the three (applied science, skills and intuition) that Li-Hua originally articulated (see Figure 6.1). The fourth form of technological knowledge is, to stay within Heidegger's domain of interest, philosophy. Adding philosophy to technological knowledge may look like an aberration, given that practitioners of technology are pragmatically minded experts unaccustomed to speculative philosophical knowledge. Yet technology and philosophy have gone hand in hand since the dawn of philosophical thinking in ancient Greece. According to Heidegger, the essence of technology always comes before the instrumental aspects of technology in the hierarchy of the 'primal truth'.[36]

In the context of biocivilisations, there is one uniquely human feature as far as technological behaviour is concerned: our perceived independence from nature, which gives us the apparent licence to use technological knowledge without any limits. For Heidegger, this is a dangerous position: techno-scientific hubris blinds us to how nature actually operates. Paradoxically, other species, whom we consider inferior, do not seem to have this problem. The unit of survival is the unit of mind (see Chapter 3), and the self-assertive tendencies of all other species are balanced by the integrative tendencies necessary to exist within the Gaian system. This natural hierarchy – imposed by Gaia's decentralised mind – is unbreakable. Yet mainstream biology interprets nature as mindless. This is a testament to the fact that integrative tendencies are generally considered superfluous in defining the human relationship to the rest of nature; modern humans have power over nature because (we believe) we have superior minds.

Heidegger's essay targets this delusion in a unique way. His first argument is that the essence of technology is neither technological nor instrumental.

This is a baffling statement. How can technology be non-technological? Yet Heidegger offers an original way out of this paradox. To capture the essence of technology, we have to establish a relationship with technology. For Heidegger, this relationship becomes obvious when we compare ancient and modern technologies. Ancient Greeks used the word *techne* for technology. *Techne* meant both technique and art. The Greeks understood arts and crafts as a process through which the world reveals itself to us. Artists and craftsmen are not the ultimate 'makers' of art and craft products. Rather, they are conduits, together with instruments and materials, in the process through which nature reveals itself to us. Heidegger calls this process *poiesis* – bringing something into being. Nature, or *physis* in Greek, is '*poiesis* in the highest sense'.[37] So the essence of ancient technology is revealing the world to us – understanding the world through making it.

Modern technologies, for the most part, do not help things come into being. Instead, they force nature to reveal itself to us as a calculable reserve of resources for human use in the form of raw materials: from planetary energetic resources such as oil, gas, water, wind, sunlight and nuclear, to genetic resources whereby the genomes of all species can be put to use for human ends. Modern technologies even enable us to go beyond our planet. According to scientists, one day we may be able to force the solar system, or even the galaxy, to reveal themselves to us as new repositories of resources for human use.[38] So the essence of modern technology is that *Homo sapiens* – thanks to a combination of the will to power and the force of technological power, in which even modern physics entraps nature in a calculable coherence of forces – becomes a self-proclaimed master of the universe. But this position is untenable. Arthur Koestler argued that the analytical power of our intellect is an unwanted evolutionary gift. The brain, according to Koestler, is the only organ we do not know how to use.[39]

Heidegger's second argument in 'The Question Concerning Technology' is equally relevant. Technology, he stated, is not the product of human activity. For Heidegger, and for the Ancient Greeks, technology is a way of understanding the world. Human beings are not in control of this process. We may create powerful machines that one day surpass human intelligence, but machine superintelligence will not improve true understanding of the world one iota. This is because nature reveals itself to us on her terms. The more we force nature into revealing herself to us, the more she hides her essence. Technology as a way of understanding the world, in the style of ancient artisans and craftsmen, is not the product of human activity; it is the product of nature's generosity, in which the truth behind natural processes reveals itself to those who are able to see it. Heidegger calls this aspect of technology *aletheia*, the Greek word for truth.[40] In Heidegger's interpretation, *aletheia* is bringing what's hidden in natural material into the light so that it shines before our eyes in the form of handiwork or art.

Lastly, Heidegger argued that modern technologies are dangerous, but not from the instrumental aspect of technology. We can live with weapons of mass destruction, AI, genetic engineering and bioterrorism. The real danger comes from failing to understand the essence of technology. If we subscribe to the modern, heavily instrumental idea of the essence of technology, in which the world reveals itself to us as a repository of raw materials for human use, our future is likely to be dystopian and replete with self-destructive dangers. In this scenario, it is not difficult to imagine the logical consequence that one day humans will become the raw material for other humans.[41]

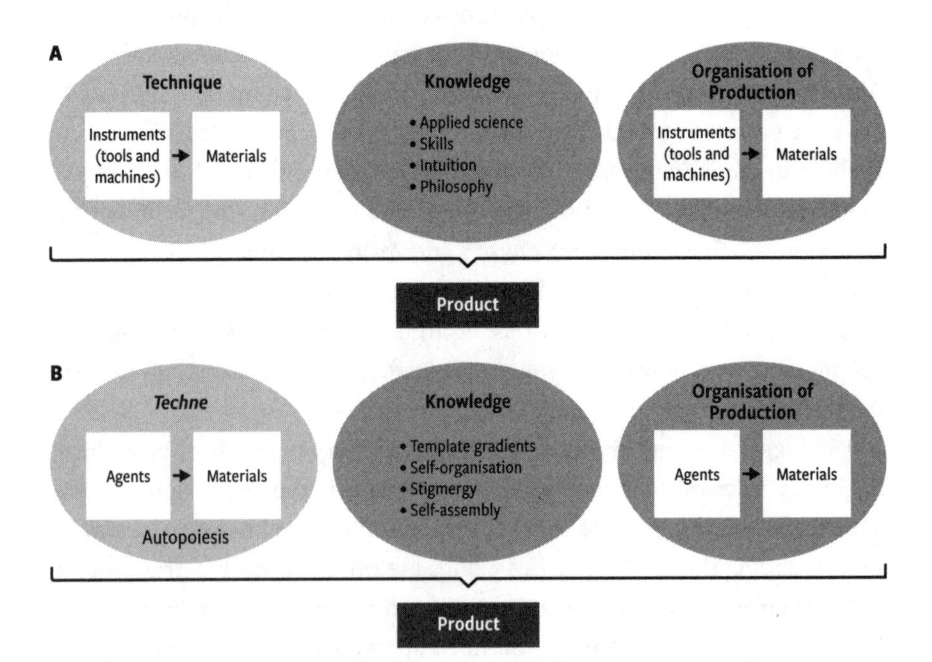

Figure 6.3 A. The knowledge component of technology enriched by philosophy. **B.** *Techne* and autopoiesis incorporated into Li-Hua's technological framework.

Heidegger sees an alternative to this self-destructive path. Exposure to technological dangers stimulates saving power, and here Heidegger relies on his favourite poet, Friedrich Hölderlin. Poetry, and art in general, may help us correct our orientation to the world towards one that offers a path to a safer and more satisfying future, by balancing master and servant tendencies, as per Koestler's concept of holon (see Figure 3.2, page 51). A good illustration of what Heidegger meant by saving power is Arcimboldo's *Flora*, which reveals to us the workings of nature, even centuries after it was painted (see 'Flora or Mona Lisa?', page 32). This is the invitation nature offers us: she reveals the truth when we submit ourselves to the essence of technology.

Heidegger's vision of technology at first seems to undermine Li-Hua's technological framework, which is heavily instrumental. However, folding philosophy into knowledge within Li-Hua's framework easily widens it to accommodate Heidegger's arguments (see Figure 6.3 A). We can further modify Li-Hua's framework by splitting technique into two parts: the *techne* and the technique, to separate modern from ancient technologies (Figure 6.3 B). In ancient technologies, technique was part of *techne*, a way of understanding the world in which humanity was an integral and subordinate part of *poiesis*, or bringing forth the world.

Engineering as Autopoiesis

Using technology as the second designator for engineering surpasses anthropocentric interpretations. We can now proceed to express the meaning of engineering in the context of biocivilisations.

The essence of engineering in nature is centred around *techne* in the context of *poiesis*. The term that faithfully captures the essence of engineering in the context of *techne* and *poiesis* is 'autopoiesis' (see Figure 6.3 B). According to Maturana and Varela, who invented the term, autopoiesis describes the essential property of nature.[42] Nature constantly reconstructs and reinvents itself because it consists of autopoietic units: organisms as self-producing, self-organising and self-maintaining agents. The integral part of autopoiesis is cognition. Organisms sense their surroundings and change them to suit their needs. In this way, organisms create environments. *Physis* is indeed the highest form of *poiesis*.

Even basic biological processes are part of autopoiesis. Gregory Bateson tried to convince his audiences at lectures and conferences that we do not see things around us mechanically, like robots or computers do. Instead, every time we open our eyes, we reconstruct the world using the natural instruments at our disposal, such as our eyes and brains.[43] Similarly, Maturana and Varela argued that we do not see the world; instead 'we live our field of vision'.[44]

But what about modern human engineering, in which technique, rather than *techne*, dominates (see Figure 6.3 A)? There is no doubt that modern engineering has brought enormous benefits to humanity, but now we have little choice but to continue using modern technologies and inventing new ones. This is a problem because technique, as the essence of modern engineering, exerts enormous destructive power that threatens human existence.[45]

How can we change our orientation to the world? This remains an open question. The conventional wisdom is to harness yet more technique-based engineering. In other words, we don't need to change our mindset – we can continue to see ourselves as engineering masters over mindless nature – we only need to improve our techniques. Heidegger's interpretation of technology, however, questions the wisdom that the needed change can be brought about by the same instruments that caused the need for such a change. In the rest of Chapter 6, I shall present the alternative through two examples of non-human engineering that instead revolve around *techne* and autopoiesis.

How Bacteria Construct Cities

Experts often use the term 'bacterial city' to describe the social organisation of biofilms, although 'city' should not be taken literally.[46] While similarities in the social organisation of humans and social insects are well recognised, bacteria are loosely organised communities that easily spread in all planetary habitats. Still, it is justifiable to use the term 'city' to capture the phenomenon in which a small number of bacterial settlers use their communication and engineering skills to construct metropolises, which permit collective responses to the environmental challenges of feeding, watering, sheltering and defence.

Powerful microscopy techniques have allowed microbiologists to study the process of biofilm construction from single cells to metropolises. The process has seven stages.[47] In stage 1, a few enthusiastic settlers – say, five bacterial cells – select a suitable hard surface to serve as a new construction site. In stage 2, the bacteria start dividing to produce chains consisting of several cells. These chains of cells start secreting extracellular polymeric substances (EPSs), which form the matrix that holds the cells together and keeps them attached to the hard surface.[48] EPSs consist mostly of a complex mixture of polysaccharides. Other molecules can include proteins, DNA and lipids.

In stage 3, chains of cells divide further to produce larger clusters of cells that resemble small villages. These clusters, consisting of about fifty cells, continue secreting EPSs. In stage 4, the small villages start merging to produce larger villages consisting of about 300 cells. The territory occupied by bacterial villages is in the region of 50 μm^2. The territory occupied by the initial settlers is usually ten times smaller. The secretion of EPSs continues. In stage 5, the villages start expanding and merging, occupying a larger territory, up to 200 μm^2.

In stage 6, the process of village mergers is completed. For the first time, a mushroom-shaped city structure emerges that consists of about 1,000 cells. This structure is also known as the microcolony. In the final stage, several microcolonies may merge to produce the bacterial metropolis, now occupying a territory of 500 μm^2. The chemical analysis of a full-blown bacterial metropolis shows that EPSs constitute up to 90% of its biomass.[49] In other words, bacteria construct metropolises consisting mostly of material secreted by their bodies. Bacteria reposition their bodies within the EPS material in an orderly fashion dictated by the processes of social interactions.

We can now map the process of bacterial city construction on the *techne*/autopoiesis framework (see Figure 6.3 B). The first thing to note is that instruments (tools and machines) are now replaced by agents. Agents, or organisms, sense their environments and respond to environmental stimuli. In other words, agents can be interpreted as sensitive body instruments. These body instruments are not machines, because the concept of the machine cannot capture the relationship between agents and their environments, as we know from Jakob von Uexküll's work and Gregory Bateson's concept of mind. Bacteria – the simplest natural agents, or autopoietic units – sense multiple parameters in their environments, from temperature and humidity gradients to surface hardness, and respond to those stimuli.

The material for building cities is EPSs secreted by bacterial bodies, but the knowledge required to construct a bacterial metropolis is still mostly unknown. It is likely to involve phenomena similar to those employed by social insects, including template gradients, self-organisation, stigmergy and self-assembly. An important aspect of construction is also constant communication (see Chapter 5). The organisation of production must include some form of division of labour. Bacteria that produce one type of EPS might position themselves at the outskirts of the city to defend it from predatory attack. Other bacteria that produce a different type of EPS might position themselves in the city centre to maintain channels through which nutrients are distributed. The product is the bacterial metropolis. The metropolis might last for a while, or it might disperse quickly to release individual bacteria that will then attempt to construct new cities.

Construction of Forest Infrastructure

Forests are the most diverse parts of the biosphere, in which the evolutionary trajectories of numerous species cross and turn into an amazing web of ecological interdependencies. The scientific term for the community of diverse organisms sharing a habitat is the 'biome'.[50] There are three different types of forest biomes: tropical, temperate and boreal. Each type occupies a different latitude and has a different climate. Tropical forests occupy the area around the equator and have warm and humid weather throughout the year. Temperate forests are located at higher latitudes and have all four seasons. Boreal forests occupy areas with low temperatures year-round and are located in Canada, Alaska and Russia.

How do forests, these wonders of biodiversity, function? For starters, their infrastructure is hidden deep underground. It consists of a complex network of roots that rely on natural engineering to make the evolutionary forest theatre tick.[51] Trees produce carbon-based nutrients through photosynthesis, but they get other nutrients such as nitrogen, phosphorus and amino acids from the soil. One strategy in the search for essential nutrients is symbiosis. Trees trade carbon-based nutrients with fungi and bacteria in return for essential non-carbon nutrients.

The first type of tree–fungus symbiosis was established 400 million years ago. Fungi known as arbuscular mycorrhizal fungi are integrated into cortical root cells. From there, the fungi project long, filamentous structures known as hyphae, which expand into the soil to make large underground networks. The second symbiosis occurred 200 million years ago. It involves a different type of fungi, known as ectomycorrhizal fungi, which does not integrate itself into root cells, but instead attaches itself to the root surface. The third type of symbiosis, involving nitrogen-fixing bacteria such as *rhizobia*, occurred 60 million years ago.[52] In this case, the roots 'capture' bacterial cells and keep them there in structures called bacteroids.

The engineering of underground infrastructure becomes apparent when scientists analyse root thickness. Trees adjust the thickness of their roots to the soil type as a survival strategy, but the beneficiaries are not only the trees but the entire biome.[53] In other words, trees integrate information from their symbiotic relationship with fungi and bacteria, as well as information about the soil, to determine how thick their roots should be in a given soil type. For example, in the nutrient-rich soil of the tropics, trees save energy by letting fungi and bacteria do the work of searching for nutrients, and they put their energy into ensuring that their roots are thick enough to optimally collect those nutrients. However, in the nutrient-poor soils typical of boreal forests and deserts, roots become thin so they can easily move through the soil in search of nutrients. These thin roots indicate that the trees are not relying on symbioses with fungi and bacteria, so much as searching independently for nutrients. This ability to adjust root thickness in search of nutrients suggests that trees are not passive automata, but active agents in their own survival.

Another element of these active ecological dynamics is trees' ability to manage symbioses with fungi and bacteria. For example, trees' relationship with nitrogen-fixing bacteria resulted in the emergence of more productive and resilient biomes capable of withstanding a range of ecological pressures.[54] Trees are intelligent cognitive agents, as Anthony Trewavas, František Baluška and their colleagues argue (see Chapter 3), and intelligence has enabled trees to spread all over the Earth; it has been estimated that forests occupy one-third of the Earth's landmass. The beneficiaries of this intelligence are not only the trees themselves, or the symbiotic fungi and bacteria, but entire biomes and all kingdoms of life. Additionally, even in an era of heavily industrialised economies, a total of 1.6 billion rural people depend on forests.[55]

Mapping the process of forest infrastructure construction by trees on the *techne*/autopoiesis framework is relatively simple (see Figure 6.3 B). Trees are intelligent agents capable of actively using their root systems and symbiotic relationships with fungi and bacteria for their benefit and the benefit of the biome. The material is the soil, fungi and bacteria. The knowledge can be characterised as a specific type of natural engineering that involves integrating knowledge from at least two sources (symbiotic microbes and soil quality) to determine the optimal root thickness. The organisation of production is spread over an evolutionary period of 400 million years. The product is an evolutionarily stable strategy for the survival of trees, which at the same time creates resilient biomes incorporating numerous species from all kingdoms of life. Trees truly support life. They are proper partners in the autopoietic physiology of the Gaian system.

The Autopoietic Planet

Bacteria, protists, fungi, plants and animals do not have human planning skills or the capacity to produce engineering blueprints. Yet they are no less successful engineers than us. If the criterion for productive natural engineering is support for life processes, modern human engineering is a failure. While trees, for example, support life processes, we destroy them. Numerous species, from all kingdoms of life, thrive in the forests thanks to the capacity of trees to engineer a sophisticated underground infrastructure for extracting essential nutrients from the soil. By comparison, our three industrial revolutions in the last 120 years – electrical, electronic-digital and cyber-physical-biological – have coincided with the decimation of animal populations worldwide, a trend that has been dubbed 'biological annihilation' by the journal of the American Academy of Sciences.[56]

The lack of human-like planning skills in other species is not a handicap; it is an advantage in the context of biocivilisations. The lack of analytical planning skills frees non-human species from forming autocratic relationships with their environments. The relationships they form with local environments are instead autopoietic.

The engineering aspect of the autopoietic Gaian system can be demonstrated by asking a simple question: where is the blueprint? In the case of ancient humanity and all other species, the blueprint is in the environment. It is hidden in the structural coupling between agents and their environments. Agents sense environmental heterogeneities, from temperature, humidity and the shapes of natural materials to pheromonal templates, sound frequencies and nutrient concentrations. These heterogeneities act as simple cues for agents to orient themselves to the world in an autopoietic way. In the case of modern humans, the blueprint is in our heads, oriented to the world by the instrumental essence of technology.

CHAPTER SEVEN

Scientists

All life is problem solving. From the amoeba to Einstein, the growth of knowledge is always the same.

KARL POPPER[1]

One of the benefits of globalisation is that in any big city, irrespective of its geographical location, you are likely to find restaurants serving food from all corners of the world. Take London, for example. During the 2018 World Cup, *Daily Telegraph* journalists identified a restaurant in the city for each of the thirty-two nations participating in the World Cup.[2] Gastronomy, it seems, is as important a global phenomenon as football. It is difficult to imagine a respectable TV channel, anywhere in the world, without a cooking programme complete with a celebrity chef. The simplest recipes, from omelettes to bread, are probed with rigour and attention to detail by TV chefs and watched with fascination by audiences around the world.

Cooking has entered yet another dimension of expertise and exactitude: the science of molecular gastronomy.[3] The founding fathers of molecular gastronomy are Nicholas Kurti and Hervé This-Benckhard. A professor at Oxford and a member of the Royal Society, Kurti had a rich career in physics. Born in Budapest to Jewish parents in 1908, he was forced to leave Hungary in 1926 because of anti-Jewish laws. He studied first in Paris and then in Berlin under Franz Simon. When Hitler rose to power, Simon and Kurti moved to the Clarendon Laboratory in Oxford, where Kurti's subsequent work in low-temperature physics was important to the Manhattan Project. He is remembered for his 1956 entry in the Guinness Book of World Records as the creator of the lowest temperature: 1 microkelvin, just a fraction of a degree above absolute zero.[4] Apart from science, Kurti was interested in music and cooking, and it was his passion for cooking that led to his contributions to the field of molecular gastronomy. In 1969 he delivered a lecture at the Royal Institution entitled 'The Physicist in the Kitchen', in which he used a microwave generator to make a reversed baked Alaska.

Kurti's enthusiasm for merging cooking with science was shared by French physical chemist Hervé This-Benckhard. Initially using the term 'molecular and physical gastronomy' to describe the new science, Kurti and This-Benckhard published an article in *Scientific American* in 1994 criticising traditional chefs for being too conservative, such that 'culinary superstitions and old wives' tales continue to flourish'.[5] They argued that science can help us understand the principles behind traditional cookery and facilitate the creation of new dishes. But Kurti and This-Benckhard rightly concluded that the scientist will never dethrone the chef. Cooking is, after all, a form of art, which only needs 'a soupçon of science'. After Kurti's death, This-Benckhard shortened the name of the new science to molecular gastronomy, and today practitioners of molecular gastronomy include many famous chefs, such as Ferran Adriá (known for the legendary El Bulli restaurant near Barcelona) and Heston Blumenthal (whose restaurant The Fat Duck is one of the most popular in Britain).

Even though cookery is often considered a trivial human activity, especially compared to fields such as art, science and engineering, Kurti's and This-Benckhard's insights have far-reaching consequences that may change our perception in the context of biology and evolution. If we interpret cookery as an activity that searches for the formula of optimal feeding, its essence becomes a widespread natural phenomenon not confined to human nutrition. An ordinary amoeba, a lowly single-cell creature, is hardly different from any chef in terms of desire for nutritional satisfaction.

Kurti and This-Benckhard introduced science into cookery, and the great philosopher of science Sir Karl Popper introduced the amoeba into the philosophy of science (Popper was bold enough to compare the amoeba's knowledge-generating skills to those of Einstein). Both Kurti's and This-Benckhard's molecular gastronomy and Popper's naturalisation of science, known in philosophical circles as evolutionary epistemology, are examples of the naturalisation of science, which I'll explore in this chapter. In brief, all living creatures, including amoebas, have their own form and practice of science.[6] (Popper's model of the naturalisation of science – evolution as periodicity of forms and functions that transcend human civilisation – will be used in the coming chapters to naturalise medicine and art and also re-evaluate communication and engineering in the context of biogenic art.)

Science and the Blob

In an important 2010 study published in the journal of the American Academy of Sciences, a group of researchers from France and Australia investigated the nutritional habits of a slime mould ameboid species called *Physarum polycephalum*, also known as the Blob.[7] The researchers discovered that the Blob has incredible gastronomical skills, no less impressive than those of celebrity chefs. But before we delve into the Blob's talents, let's explore what sort of organism it is.

The Blob is a protist, which is a single-cell organism formed by bacterial–archaeal mergers (see Chapter 5). It has a complex life cycle. In a wet environment, *Physarum polycephalum* is a single cell that reproduces asexually without cell division by multiplying its nuclei. The result is a glistening yellow biogenic mass that scientists refer to as plasmodium, which can occupy a space of several hundred square centimetres. In other words, you can see the Blob with the naked eye.

For scientists, the Blob is a wonderful experimental toy. There are only two requirements to play with it: the first is a wet Petri dish, a surface on which the Blob will live contentedly; the second is food. The easiest and cheapest way to provide a meal for the Blob is to share your muesli with it by placing a few oat flakes on the wet surface of the Petri dish. Then you can watch the marvellous spectacle of the Blob's growth and movement. (If you do not want to experiment with the Blob yourself, I recommend watching it on YouTube. Just search '*Physarum polycephalum*' and you will find a series of videos that will delight you for hours.) In particular, you will see the emergence of a spectacular network of yellow tubules that move quickly towards the oat flakes and consume them with gusto.

While playing with the Blob, scientists have learnt important tricks to help them study the Blob's behaviour. For example, you can remove food and see what happens when it starves. Or you can offer the Blob more than one food source and then analyse its different responses, as French and Australian scientists did when they offered the Blob not two or three different 'meals' but thirty-five, each with a different source and mix of nutrients.[8] In the first experiment, the Blob was offered a choice of two meals. In the second experiment, the Blob was placed in the centre of the Petri dish and offered a choice of eleven meals, positioned in a circle around the periphery. The scientists also varied the combination of meals, either in pairs or in groups of eleven meals.

The results of this research were staggering. The Blob was able to identify and preferentially consume meals that contained concentrations of nutrients optimal for its growth and behaviour. Scientists commented that this brainless creature is as smart as we are, at least in a sense.[9] Its capacity to identify a diet that contains the optimal ratio of proteins and carbohydrates is a sign of intelligence despite the lack of a brain.

This is to say that Karl Popper was not exaggerating when he compared the Blob to Albert Einstein. What they have in common, according to Popper, is the acquisition of knowledge driven by the need to solve problems.[10] The Blob solves the problem of optimal feeding and nutrition; Einstein solved important problems in science. That said, Popper also identified a crucial difference between Einstein and the Blob. In the trial-and-error problem-solving process, the Blob will die if it makes an error. Einstein, on the other hand, knew that errors were essential in order to refine his knowledge, so he deliberately designed new trials to detect new errors. Scientists use trial and error to identify and eliminate errors and thereby refine scientific theories. To paraphrase the great John Keats, error is a highway to success.

Science – in the form of trial-and-error exercises directed at problem-solving – is a widespread natural phenomenon not confined to human civilisation. We can conclude that Popper was correct in his understanding of science as trial and error aimed at the solving of problems from further examples of the Blob's intelligence. For example, scientists realised that the tubular pseudopods produced by the Blob are a sort of distributed information processor. Cytoplasm streams rhythmically back and forth through the tubules and, in the process, the Blob detects nutrients and chemical signals in the environment. The constant assessment of the environment guides the network of tubules to move in a particular direction.

This is how the Blob quickly learns the shortest route to food when scientists place the Blob in a maze.[11] In fact, the Blob is so skilled that in an experiment designed by Japanese scientists, it was able to reproduce the greater Tokyo railway network.[12] The scientists placed an oat flake at each point on a map to represent towns on the outskirts of Tokyo. They then left the Blob to do its thing. The Blob's pseudopods visited all the oat-flake 'stations', established a series of 'corridors' linking them and even tested the network to probe its efficiency. It eliminated many corridors as suboptimal. At the end of twenty-four hours, it had established a large network of tubules and then refined it to several routes that allowed for optimal feeding. What was most remarkable is that the network the Blob produced very closely resembled the greater Tokyo railway network. The Blob's capacity to reproduce this network prompted the scientists to develop a mathematical model for constructing other adaptive networks based on the Blob's scientific intelligence.

So Popper was correct in another sense. If Einstein and the Blob use the same principles to generate knowledge – problem-solving through trial and error – they can also exchange knowledge with each other to refine their strategies. The universality of knowledge creation is one of the principles behind the philosophical discipline of evolutionary epistemology.

Evolutionary Epistemology

The philosophical discipline of evolutionary epistemology, founded by an American psychologist, social scientist and philosopher named Donald T. Campbell (and supported by Karl Popper), is based on three principles: (1) evolution is a cognitive process, (2) organisms are knowledge systems and (3) all forms of knowledge share certain features.[13]

Popper's creativity in linking Einstein and the Blob nicely summarises all three principles behind evolutionary epistemology. The first principle (evolution as a cognitive process) was described by Popper as problem-solving through trial and error. The Blob is an excellent example of natural ingenuity at the heart of evolutionary epistemology, but there are many more. In Chapter 5, we saw how the protist *Spirostomum ambiguum* solves the problem of predation by inventing hydrodynamic waves as a mode of detraction. We also saw how trees solve the problem of nutrition and how those solutions benefit many other species, as well as how trees defend themselves against leafcutter ants by releasing VOCs. This simple two-way relationship between trees and ants is the basis for problem-solving within ecological relationships that involve thousands of species. We've also seen how bacteria solve the problem of turning a dead planet into a living one (see 'The Internet of Living Things', page 16).

If the problem is denoted as P, its solution, or tentative solution (TS), and the error elimination (EE), the 'fundamental evolutionary sequence of events', can be described in the following way:

$$P \rightarrow TS \rightarrow EE \rightarrow P.$$ [14]

However, the solution to the problem, for example, of how trees extract phosphate from the soil (through symbiosis with fungi), always creates a new problem. The problem of the need for nitrogen is then solved through symbiosis with nitrogen-fixing bacteria (see Chapter 6), so the second problem is always different from the first one. Therefore, a more accurate expression would be:

$$P1 \rightarrow TS \rightarrow EE \rightarrow P2.$$

Even this formula is inaccurate, however, because it ignores the multiplicity of tentative solutions or the multiplicity of trials. A third, more accurate, expression is:

$$P1 \rightarrow (TS1, TS2, ..., TSn) \rightarrow EE \rightarrow P2.$$

Popper further interpreted his formula in the context of neo-Darwinism. The key problem in evolution is survival. The multiplicity of TSs to the problem of survival, according to neo-Darwinism, is the variety of mutations. However, we have seen that neo-Darwinism only accounts for one side of the biological coin. Natural variation of mutations, as a form of biological problem-solving, is limited. Neo-Darwinism must be balanced with Lamarckism, which better accounts for organism–environment interactions and epigenetics, a form of genetics that places less emphasis on the determinism of gene mutations.[15] Contemporary evolutionary epistemologists often integrate Darwinism and Lamarckism for a more complete understanding of cognition as a biological phenomenon.[16]

The second principle of evolutionary epistemology (organisms as knowledge systems) is best demonstrated by continuing to use the Blob as an example. The Blob's cognitive skills, in the absence of a brain, are staggering. They range from finding the shortest route to food in a maze to reproducing a complicated urban transportation network. Numerous experiments have shown that genetic mutations do not play any role in the Blob's ingenuity. For example, the Blob can replicate Tokyo's railway network within a period of twenty-four hours – too short a time for a mutation to emerge and act as the variant that leads to the solution.[17] Instead, the Blob's pseudopods act as distributed information processors that constantly assess the environment and enable the Blob to eliminate possibilities that don't represent optimal solutions.

In this way, rigid genetic determinism is replaced by a far more flexible problem-solving tool based on the concept of biological information. Gregory Bateson elaborated the crucial difference between information in the biological sense and information in the engineering or physicalist sense.[18] Organisms are constantly searching for meaning. Biological information, or the 'difference that makes a difference', is a semantic concept. Life is a meaning-making phenomenon.[19] On the other hand, machines are programmed to 'interpret' information as a message without meaning. This is a purely syntactic concept. Natural learning and machine learning are fundamentally different processes.[20]

The third principle of evolutionary epistemology (all forms of knowledge share certain features) implies that natural knowledge is interchangeable. In practical terms, this means that a solution to the problem of the organism–environment interaction, identified at one level of the biological hierarchy, can be copied to another. For example, flight and vision emerged periodically at different evolutionary times and in different phyla. The emergence of the same biological phenomenon at different evolutionary periods, and in different species, is known as evolutionary convergence or biological periodicity (see 'Evolutionary Retrodiction', page 71). Donald T. Campbell described how 'the language of the bees', which was invented millions of years ago, is analogous to the language of humans, a much more recent evolutionary phenomenon.[21] Similarly, Eshel Ben-Jacob argued that the oldest evolutionary form of communication is 'bacterial language'.[22] All natural languages – bee language, bacterial language, plant language and human language – share the same principles of information exchange, as we know from contemporary biosemiotics (see 'A Short History of Biosemiotics', page 86).

Human engineering benefits from copying solutions that other organisms invented to solve their own problems. This is known as biomimetics (see Chapter 6). Here's one example: if you have a winter coat, you likely have some sort of Velcro fastener on it. Velcro was invented by a Swiss engineer named George de Mestral when he noticed that any time he went walking in the Alps with his dog, both he and his dog returned home with burdock seeds stuck to them. Being a curious engineer, de Mestral examined the seeds under a microscope and saw they had numerous hooks that grabbed onto the fibres of his socks and his dog's fur. What he was noticing was a solution that the plant *Arctium* (aka burdock) invented to solve the problem of seed dispersal.[23]

De Mestral surmised that this gripping device might be a good alternative to zippers, which often jammed, but his idea of a new hook-and-loop fastener that copied burdock's method of seed dispersal was initially ridiculed. De Mestral also had technical problems with identifying the best type of textile for the fastener. After experimenting with different materials, he discovered that a strip of nylon sewn under infrared light contains tough hooks that attach firmly and reversibly to a strip of fabric with less tough hooks. De Mestral patented the idea under the name Velcro – *velrous* (velvet) and *crochet* (hook) in French – and founded a company that he later sold to American investors.

What Is Science?

Even though all organisms are 'scientists' in a limited and metaphoric sense (provided we accept Popper's thesis that life is an endless process of problem-solving based on the method of trial and error), the question remains as to how evolutionary epistemology interprets and understands science. Here, we have a range of opinions that reflect the different scientific and philosophical assumptions of evolutionary epistemologists.

For Karl Popper, the growth of scientific knowledge was reminiscent of evolution itself. He thought scientific hypotheses represent blind trials exposed to the filter of selection, which eliminates non-successful trials.[24] This view emulates the basic tenets of neo-Darwinism. Scientific hypotheses are equivalent to variations in DNA sequences at specific genetic loci, or mutations that specify a biological function encoded in the protein structure. Natural selection eliminates most mutations but retains and propagates those that lead to new biological functions or to the improvement of existing functions. In the same manner, scientific hypotheses are selectively eliminated if they are not compatible with scientific theories, which in Popper's metaphor mimic biological function. Popper's reliance on the basic tenets of neo-Darwinism was not surprising. At the time he was developing his ideas in the 1970s, neo-Darwinism was at the peak of its popularity and had essentially monopolised biological theory.

Popper's colleague Donald T. Campbell shared this selectionist view. In addition, Campbell proposed a hierarchy of evolutionary knowledge consisting of ten levels.[25] Level 9 was occupied by human culture, surpassed only by science, which, as a product of human culture, occupied level 10. Neither Popper nor Campbell, however, ventured outside Darwinism and neo-Darwinism in the search for principles behind evolutionary knowledge. This is a problem for two reasons. First, as I argued in Chapters 3 and 5, Darwinism – the survival of the fittest – is only one side of the biological coin. Lamarckism – the survival of the ecologically most intelligent (see 'Challenging Anthropocentrism', page 80) – plays as important a role in evolution as Darwinism. Second, placing (human) science at the top of the pyramid of evolutionary knowledge disregards the discrepancy between the timeline of evolution and the timeline of *Homo sapiens* (see Figure 3.1, page 48).

Some evolutionary epistemologists have acknowledged the problem of neo-Darwinian exclusivism. The American evolutionary biologist Richard Lewontin challenged genetic determinism by emphasising organism–environment interactions.[26] Not only do organisms actively create environments through the process of natural learning, but the environments actually become their learning partners. We see this clearly in the case of the forest ecological theatre, in which evolutionary trajectories of numerous species cross and intertwine with each other, leading to new biological functions. For example, in the forest underground network (see 'Kingdoms Fungi, Plantae and Animalia', page 97) there is no hard distinction between organisms and environments. Trees, when we analyse them in the context of evolutionary knowledge, turn into dual entities similar to Koestler's holons. Their bodily structure – from roots to leaves – is the true self, or the true organism, with clear boundaries between itself and the environment. But trees also have an external extension separate from their true self, in the form of symbiotic fungi associated with their roots. This extension – the fungal underground network – is part of the environment and yet it remains functionally integrated with trees to the extent that trees use it for communication and food transport. Trees are not unique. All organisms other than bacteria are two-faced entities – part organism, part environment – like the Roman god Janus, which was the basis for Koestler's concept of the holon. Organisms and environments, in other words, are relative concepts (see 'Flora or Mona Lisa?', page 32).

The lack of a functional boundary between organisms and their environments is further underscored by the concept of universal symbiogenesis proposed by the theoretical physicist Freeman Dyson.[27] Dyson argued that biological symbiogenesis is an extension of the cosmic symbiogenesis that characterises the formation of galaxies and other cosmic bodies. Dyson acknowledged Lynn Margulis as the key thinker behind the concept of symbiogenesis after Margulis published her seminal 1967 paper 'On the origin of mitosing cells', in which she argued, against the tide of scientific opinion, that all eukaryotic organisms are ecological collectives.[28] We, like all other animals, are coevolved microbial communities.

The concept of universal symbiogenesis challenges Campbell's hierarchy of evolutionary knowledge. If there is a 'top' of the biological hierarchy, it is Gaia. As a symbiotic system, Gaia is a distributed problem-solver that integrates all its components in the search for evolutionary knowledge at the system level, and it is organised in such a way that the components of the system are subordinate to the system itself. Suggesting that humanity (a component) surpasses the knowledge of the system itself (Gaia) is akin to violating natural law. It would be like suggesting an evolutionary epistemology that violated the second law of thermodynamics.

How can we reconcile science and Gaia? I propose two answers to this question. The first reflects the pragmatism of scientists and philosophers of science. Science does not search for absolute truth. In the relationship between Gaia and science, Gaia is absolute (in a limited sense): a system that controls its components. Science, on the other hand, looks for the logic of the component as it searches for its place within the system. As perpetual trial-and-error exercises, science must always challenge itself; theories are not scientific if they cannot be disproved. The art of disproving theories, according to Richard Feynman, always includes the courage of challenging previous generations of scientists (see 'Humanist Pride', page 63). The solutions that science gives us are always temporary.

My second response to the question of how we can reconcile science and Gaia is focused on the structure of science. It challenges potential errors emanating from it. For example, the belief that humanity (a component of the system) can replace Gaia (the system) as the most dominant environmental force on Earth is almost certainly erroneous.[29] The key thinker who challenged the structure of science was Robert Rosen, who argued that science must do a U-turn to understand life: it must replace the primacy of physics with the primacy of biology. Reductionist-mechanistic approaches might work well in physics, but in biology these approaches are limited because biological systems are complex. The key feature of complex systems is that they remain more complex than our models of them.[30]

To appreciate Rosen's thinking, it is essential to understand the difference between organisms and machines. The structure and organisation of machines are fully computable. This means that our mathematical models of machines fully describe them. By contrast, the structure and organisation of organisms are non-computable. Our mathematical models always lag behind the complexity of biological systems.

Scientists are divided on whether science is capable of crossing the Rubicon and understanding life purely through computation. In 2010, scientists created the first bacterium with an artificial genome synthesised by a computer; they named it *Mycoplasma laboratorium*.[31] The same group of scientists orchestrated a media campaign to promote the idea of synthetic biology. In response, J. Scott Turner argued that the artificial genome would not have worked without already existing life in the form of the bacterial cell from which the genome was removed.[32] Nobel Prize–winning physicist Roger Penrose similarly argued that consciousness, whatever it may be, is not based on computation (see 'Prejudice No. 3: Sentience and Consciousness', page 57). Consciousness is the key feature of life, from bacteria to humans.[33]

So what science 'is' depends on our attitude or orientation to the world. The current attitude of the scientific thought collective is that physics rules. In the words of James Watson: 'There is only one science, physics; everything else is social work.'[34] This suggests that reductionism and mechanism are incontrovertible tools in our understanding of the world, including the living world. But as I argued in Chapter 2, life is not mechanism. To fully understand life, we need to rethink the structure of science. What if biology, not physics, is first in the hierarchy of the sciences?

Gaian Science?

To further naturalise the concept of science, we have to ask a question that neo-Darwinists dismiss as unscientific: does Gaia have her own form of science in the naturalised Popperian sense? This question is not easy to answer because of the clash between how neo-Darwinists understand life and how proponents of the Gaia hypothesis, in particular Lynn Margulis, viewed life in its global incarnation.

The Gaia hypothesis had a troubled birth. Immediately after James Lovelock and Lynn Margulis articulated the hypothesis, neo-Darwinists dismissed it as unscientific. Leading neo-Darwinists of the day, including Richard Dawkins and W. Ford Doolittle, poured scorn on Lovelock and Margulis' arguments that the biosphere is a planetary-scale system capable of self-regulation and maintaining its own structure. Five decades later, things have changed – somewhat. The Gaia hypothesis is now integrated into the discipline known as Earth system science, and the contributions of Lovelock and Margulis have been acknowledged, but the tension remains between the basic tenets of the Gaia hypothesis, encapsulated in the concept of autopoiesis, and the dominant cybernetic approach that interprets life as a system amenable to interventions based on mechanistic solutions that deny Gaia's autonomy.[35]

Another important change occurred. A former critic of the Gaia hypothesis, W. Ford Doolittle, has recently attempted to reconcile it with Darwinism. Doolittle, a scientist and philosopher of biology, is aware of the pitfalls. In a paper entitled 'Darwinizing Gaia', he states that he has warmed up to the Gaia hypothesis, relative to his earlier uncompromising position.[36] In a nice personal touch (unusual for scientific papers), he says: 'I dedicate the exercise to Lynn, who would no doubt have thought it superfluous.'[37] Whether Dr Doolittle intended it or not, his attempt to Darwinise Gaia reflects the tension between Darwinism and Lamarckism. Let's examine both the cybernetic versus autopoietic tension, as well as the Darwinism versus Lamarckism tension.

Cybernetics versus Autopoiesis

The key assumption behind autopoiesis is that organisms are autonomous and self-producing biological systems capable of anticipation.[38] The distinction between organisms and their environments has been blurred ever since the first endosymbiotic event, when an archaeon and a bacterium merged into a eukaryotic cell. The cell became both an organism and an environment at the same time, reminiscent of Koestler's two-faced holon. Further complexification into more complex eukaryotic cells, and later into multicell organisms, led to new forms of cognition and new forms of organisms and environments. Gaia, as the system that integrates all organisms and all environments, was there from the outset, and was changing all the time.

Autopoiesis is rejected by most Earth system scientists, who typically interpret organisms as passive biological systems shaped by the actions of genes, which turn organisms into 'lumbering robots', in the words of Richard Dawkins.[39] The robots are selected by the environments only if they fit the environment's features. Thus, biological innovation is the consequence of mechanical action. The multiplicity of robotic forms, shaped by changes in genotype or random mutation, serves as the repertoire of biological forms. A limited number of these forms have a chance to survive in particular environmental conditions. Given that organisms are organic robots, their populations are shaped by positive and negative feedback loops when they encounter other populations and other environments. Feedback loops are like mechanical ropes that perfectly fit the input–output cybernetic models. Consequently, Gaia – the sum of all organisms, environments and feedback loops acting in unison – can be described faithfully by cybernetic input–output models. Not only that, Gaia can also be manipulated and modified in the programme of Earth stewardship by implementing solutions emanating from cybernetic models.[40]

The major objection of Earth system science (and its cybernetic approach) to autopoiesis as an explanatory model behind Gaia is a perceived lack of scientific rigour and emphasis on descriptive rather than quantitative approaches.[41] This objection is unjustified. Robert Rosen developed a strong mathematical background for his theory of anticipatory systems and organisms as cognitive agents.[42] Autopoiesis and relational biology (see Chapter 3) are closely linked.[43] In addition, Sergio Rubin and colleagues have recently provided a proof of concept for the formalisation of autopoiesis in a mathematical sense.[44] Therefore, autopoiesis research has a clear scientific basis.

Darwinism versus Lamarckism

W. Ford Doolittle is both a high-calibre scientist and a philosopher with elements of artistic flair in his writings. He uses magnificent analogies to describe evolutionary encounters between organisms and environments, to the point where he can also be described as an artist.[45]

Dr Doolittle's attempt to Darwinise Gaia reflects his intellectual versatility, and he describes it in the language of a movie title (but inversely) – 'It's the song, not the singer' – to explain his vision of a Darwinian Gaia.[46] He likens organisms in the microbial world (individual bacteria and archaea) to singers. Songs represent extended phenotypes of microbial singers (biogeochemical cycles, or metabolic pathways, regulated by global feedback mechanisms). When all organisms join voices together, the biosphere becomes a single complex concert. Natural selection acts to 'amplify and diversify' singing capacities. Groups of microbes, which may be taxonomically different, form associations or 'guilds' based on their functional similarities (sharing a core suite of functional genes). Guilds may come and go, but they will always turn up when required. In other words, melodies can be downplayed, but they never disappear completely.[47]

Dr Doolittle also challenges the meaning of reproduction in the microbial world. He argues that what is selected for is persistence: the recurrence of biogeochemical cycles over time. Finally, he contends with the key objection of neo-Darwinists to the Gaia hypothesis: that Gaia cannot reproduce and is therefore not subject to natural selection. His solutions to 'Gaia's population problem' range from replacing reproduction (populations matter) with persistence (biogeochemical cycles matter) to downgrading reproduction to limited value (the entire Gaian repertoire of species has a common descendant).

Some analogies employed by Dr Doolittle can also describe autopoietic Gaia (provided the right arguments are used to explain the analogy) – for example, Gaia as a unity of all singers and songs, in the form of a complex choral collective. However, the attempt to Darwinise Gaia encounters problems on at least three grounds. First, microbes do not behave according to Darwinian rules. As I explained in Chapter 5, the microbial world operates on Lamarckian principles. Darwinising microbes means recasting evolution, indeed forcing it to conform to the rules of animal and plant reproductive behaviour, which are not present in microbial populations. Lynn Margulis was not alone in voicing her objections to downplaying the microbial autonomy within the Gaian system. Carl Woese, Sorin Sonea and, more recently, Eugene V. Koonin argued in favour of Lamarckian rules in the microbial world (see 'Kingdom Bacteria', page 90). Songs and singers have different definitions depending on (Darwinian or Lamarckian) context.

Second, the concept of symbiogenesis is largely ignored by Doolittle's Darwinised Gaia. This means that novelty in the biological world is exclusively the product of natural selection – songs can only be better tuned by selecting singers better suited to certain environments but not created *de novo*. However, natural selection – a filter that lets some organisms persist – is only an editor, not an author or creator of new forms. This creator is likely to be symbiogenesis. And here we need not rely only on Lynn Margulis. Freeman Dyson developed the concept of universal symbiogenesis to describe symbiogenesis that goes far beyond the biological world and may even act as the creator of novelty on a cosmic scale.[48]

Third, we have to ask a question about the musical composition, or the actual nature of the notes, in Dr Doolittle's analogies. As far as I can judge, the nature of the notes can be described as follows: a core suite of functional notes that make melodies is genes. Songs, or biogeochemical cycles, are phenotypes. The choral singing is tuned through feedback loops. However, genes, phenotypes and feedback loops represent the nuts and bolts of the mechanical vision of life. Organisms, on the other hand, are self-producing, autonomous agents – autopoietic units in the autopoietic Gaian system, and they extend far beyond mechanical models.

We can now go back to the question expressed by the heading: does the Gaian system practise science in the Popperian sense? Yes, it does. Gaia is an autonomous biological system that has persisted for the last three billion years, using a decentralised form of trial-and-error problem-solving (see Figure 2.1, page 44). My term for this Gaian science is 'hyperthought' (see Chapter 2) – the process of bringing together the cognitive actions of all natural agents in a distributed and decentralised way, in the search for new and unimagined cognitive worlds. I do like Dr Doolittle's musical analogies, but I think Gaia can be better described as a sculptor sculpting itself using a decentralised and distributed mind: hyperthought.

Gaian Science versus Human Science

Strangely, Dr Doolittle ended his paper that described Darwinising Gaia by pitting human science against Gaian science. He wrote: 'And as we enter the Anthropocene, even those who see Gaia as an organism must admit that she can be killed.'[49] The notion that we humans can kill Gaia is a delusion dealt with by Lynn Margulis in her book *Symbiotic Planet*.[50] We humans can perturb the Gaian system; there is no doubt about it. The anthropogenic ecological crisis in the form of biological annihilation and climate change is a clear sign of our impact on Gaia. But in the process of anthropogenic disturbance, we have exposed ourselves to the mercy of Gaia's response. Gaian hyperthought will stabilise the system with or without us.

It is essential to remember that Gaia has survived the rigours of turning a dead planet into a living one. The Oxygen Catastrophe was an endogenous event caused by life – cyanobacteria producing oxygen – that was probably far worse than anthropogenic pollution. Gaia has also survived five major extinction events in the last 600 million years.[51] The K–Pg extinction event 66 million years ago was likely caused by the impact of a comet or asteroid 10 to 15 kilometres in size. The energy it released was far greater than that of all nuclear weapons combined.[52] Yet Gaia essentially swallowed the violence of the K–Pg extinction event. Even if we wanted to kill Gaia using powerful scientific technologies like nuclear weapons, we would fail.[53]

The possibility of humans destroying the Earth is a fallacy promoted by at least three lines of thinking. First, modern education trains students to view our species as the most intelligent, and thus the most powerful, in the history of life. Second, modern science promotes a mechanistic outlook. Physics is the most fundamental science; given that biology is derived from physics and physics is based on reductionism and mechanism, biology is based on reductionism and mechanism by default. Third, neo-Darwinism, the most dominant ideological force in biology, is allied with the mechanistic outlook of physics. Organisms are interpreted as organic machines controlled by the nuts and bolts of genes and environments.

While we can refute the fallacy that humans are capable of destroying life on Earth by deconstructing the three assumptions on which it is based, we cannot ignore the importance of the clash between how modern science is practised and how naturalised Gaian science operates. A point of contention has recently emerged. Using chemical organisation theory and the Deficiency Zero Theorem, researcher Sergio Rubin and his colleagues have demonstrated that Gaia is an autonomous system capable of anticipation in the manner Robert Rosen described. Autopoiesis is 'a more fundamental state of affairs than feedback self-regulation'.[54] The key feature of autopoietic systems is the distinction between structure and organisation. The structure may change as long as the self-producing autopoietic organisation of biological systems, including Gaia, is preserved. This means that the Gaian system may undergo dramatic structural changes, such as extinction events and other tipping points, without any impact on her autopoietic organisation. The process of self-sculpting continues unabated.

The point of contention between Gaian science and human science revolves around the concept of Earth stewardship.[55] Climate change models, for example, are based on feedback control theory. As part of geoengineering efforts to reduce global warming, one idea is that sulphate aerosols could be released into the atmosphere to increase the planetary albedo, thereby cooling down the planet.[56] This geoengineering plan could work, but only if Gaia weren't an autopoietic system. Gaia's response to the perturbation caused by such geoengineering would result in a range of unpredictable structural changes aimed at preserving her autopoietic nature. The bottom line is that autopoietic systems cannot be modelled by feedback control theory. Living systems are not machines that can be modified by input–output cybernetic models.

Yet modern science is entirely based on mechanistic cybernetic models, and largely ignorant of autopoiesis. This is reflected, for example, in how modern states force science to demonstrate 'relevance to wealth creation'.[57] For example, the EU Horizon 2020 programme is a science-funding programme worth €80 billion that has been dubbed a 'financial instrument' by EU bureaucrats for turning Europe into a globally competitive economic player.[58] Likewise with the Horizon Europe programme, worth €95.5 billion. A UNESCO Science Report, *The Race Against Time for Smarter Development*, reveals a global competition for innovation. Global expenditure on science was $1.767 billion in 2018, with the biggest investors also being the world's two leading economies: the United States and China.[59]

The Future of Science

But what does the future hold for science? We have at least two options. One is to continue with our mechanised science and cybernetic input–output models. Leading scientists and politicians value models based on the structure of science dominated by physics. These models, they argue, will help us overcome the ecological crisis by applying various forms of geoengineering as remedies against past ecological errors of our own making. Would this work? Even if mechanised science shows some success in tackling ecological challenges, putting all your eggs in one basket is rarely wise.

The second option is to distribute our eggs in at least two baskets. In other words, we must have alternatives to current cybernetic models. The autopoietic models must be given a chance. This does not require eliminating cybernetic models. We can simply use autopoietic models as backup in case cybernetic models fail. The values of autopoietic thinking have been articulated by Fritjof Capra and Pier Luigi Luisi in *The Systems View of Life*.[60] In addition, we now have the proof of concept that autopoiesis is a proper research programme, thanks to Sergio Rubin and colleagues. By developing autopoietic science further, we will enhance our understanding of the biosphere from an angle that mainstream science has underestimated. This would make our future responses to Gaian actions richer in scope and more in tune with her autopoietic organisation.

One thing is clear, however: Gaia will test our intelligence and the intelligence of our science by probing whether our responses to ecological and climate changes are aligned with her autopoietic organisation. The Anthropocene, after all, can never replace Gaia as the dominant ecological force on Earth.

CHAPTER EIGHT

Doctors

'Disease' is generally held to refer to any condition that literally causes 'dis-ease' or 'lack of ease' in an area of the body or the body as a whole.

JULIAN REISS and RACHEL A. ANKENY[1]

In Chapter 7, we saw how science can be naturalised. The discipline of evolutionary epistemology enables us to view all organisms as scientists (in a limited and metaphoric sense) – that is, if we accept Karl Popper's proposition that organisms learn about their environments and themselves through the endless process of trial and error. In this chapter, I'll similarly attempt to naturalise medicine. The job of naturalising medicine – the notion that all organisms are doctors (again in a limited and metaphoric sense) – is more difficult because there are neither clear-cut research programmes aimed at the naturalisation of medicine, nor supporters of such programmes of the calibre of Karl Popper in the context of science. Yet the naturalisation of medicine is possible, and justifiable, as you will see.

Before I embark on the route of naturalising medicine, here is the reason for it: the aeons-long domination of bacteria on Earth (see Figure 3.1, page 48) has been marked by two medical phenomena: immunology and epidemiology. Bacteria protect themselves against viruses on a single-cell level (immunity); they also fight viral infections on the global scale when the bacteriosphere encounters the virosphere (epidemiology).[2]

Some scientists might object to aggrandising microbes as both vehicles (bacteria) and drivers (viruses) of evolution.[3] For example, futurists argue that in the post-biological evolution that is confidently anticipated by some leading scientists, microbes may become irrelevant. Even people may become irrelevant if machine superintelligence and AI take over the planet, and beyond, one day.[4]

Post-biological evolution, however, is not a concept without problems. The main one is the idealisation of the evolutionary power of the human species and its technologies (see 'Prejudice No. 4: Post-Biological Evolution', page 61). Are we perhaps victims of our own self-aggrandising tendencies? The education system has instilled in us the view that *Homo sapiens* is the most intelligent and powerful species in the history of life. The concept of the Anthropocene – 'humans have replaced nature as the dominant environmental force on Earth' – is the measure of our self-confidence.[5] But when we see ourselves in the evolutionary mirror, human self-confidence becomes misplaced. The more we investigate microbes, the more we become aware of our insignificance on the evolutionary scale.

The most recent discovery that questions our misplaced self-confidence is the new tree of life. The tree of life is an important model in the biological sciences and evolutionary studies, used, amongst others, by Charles Darwin. It describes the diversity of life on Earth, and the interconnectedness of all life forms through common descent from a universal ancestor.[6]

In an attempt to update the existing tree of life, a group of scientists combined advances in genomics and computation with a random sampling of planetary environments. The final result took scientists by surprise. The new tree of life is dominated by bacterial diversification.[7] The other two domains of life, Archaea and Eukarya, are much less diversified; they occupy smaller branches of the tree of life. Microbiologist Dr Letitia Wilkins commented: 'I still believe in revolutions. And sometimes they just happen, almost unnoticed. One such revolution happened on a boring 11th of April 2016.'[8]

11 April 2016 was the date of publication of the new tree of life. Astrobiologists Charles Lineweaver and Aditya Chopra argued that the new tree of life is nothing short of 'the overview effect': a cognitive shift caused by a new observation that shatters our 'dogmatic common sense'.[9] One such observation was an iconic photograph of Earth taken on 14 February 1990 by NASA's *Voyager 1* from a distance of 3.7 billion miles from Earth. The photograph inspired Carl Sagan to write a powerful passage dedicated to our planet, a 'pale blue dot', insignificant and lost within the grandeur of the cosmos.[10] Sagan's pale blue dot in turn inspired Lineweaver and Chopra to express the sense of *Homo sapiens* being lost in the grandeur of biological diversity. They wrote: 'In this tree of life, ours is a small voice in a chorus of hundreds of millions of voices. We often think we are the

soloist, but in the tree of all life, we are a small new voice in an ancient choir of prokaryotes.'[11]

The starting point in naturalising medicine is to acknowledge the primacy of microbes in all aspects of life, including the self-preservation of body integrity – a quasi-medical practice in the form of bacterial immunity. This same primitive medicine also exists amongst non-microbial life forms, including plants and animals. The existence of immune systems in plants and animals, combined with widespread viral pandemics in both kingdoms of life, highlight the importance of self-preservation, or quasi-medical practices in evolution. These practices become more sophisticated with the emergence of the brain. As a result, many animal species show capacities to self-medicate. This opens a route for us to search for the roots of human medicine – the most sophisticated form of body and mind preservation – in the natural world.

In the next two sections, I shall outline the relationship between medicine and evolution, and principles of self-preservation. This will be followed by descriptions of bacterial, animal and plant immunities, and also animal self-medication. In the end, I'll reassess human medicine in the context of naturalised medicine.

Medicine and Evolution

As philosophers of medicine Julian Reiss and Rachel A. Ankeny describe, one way to think of illness is as 'dis-ease' or 'lack of ease' (see epigraph, page 139). In the state of sickness, everything in your body becomes uneasy. Depending on the disease, you may have difficulty walking, talking, breathing, eating, thinking, seeing, making love, enjoying food, going to work and doing other things that are relatively easy for a healthy person. When you are sick, you feel the lack of ease – the ease with which your disease-free body engages in everyday activities.

Disease is part of life. It happens every day, in different forms and in numerous communities of humans, animals and plants all over the planet. If you are healthy, you should be able to recover from most diseases. Your body is a flexible system. It can overcome 'dis-ease' and return to a healthy state – a state of the body and mind that can be characterised as the 'ease of existence', or a 'dis-ease'-free existence. French doctor and philosopher Georges Canguilhem described the ease of existence in the following way: 'To be in good health is being able to fall sick and recover; it is a biological luxury.'[12]

We become aware of this biological luxury – the ease of existence – only when we fall sick. But the biological luxury is also dynamic. As we become older, biological luxury becomes more elusive. Our body gradually loses the capacity to recover after sickness, or even from everyday existence. Ageing is part of life's dynamism, ending with the loss of the capacity to recover. The Serbian writer Svetislav Basara, who was influenced by Georges Canguilhem, reminds us that death is part of life: 'You do not die because you are sick but because you are alive.'[13]

Ease of existence (a state of health), lack of ease (a state of 'dis-ease'), the capacity to recover and the processes of ageing and death are all parts of the dynamics of life – not only human life, but the lives of all organisms from bacteria to animals. However, there is another part of the dynamics of life that remains elusive to us humans because we do not experience it, or if we do, it is not physical. How can we recognise an important part of the dynamics of life that we cannot experience?

One way is to investigate the life cycles of other species and identify that which is absent from the human life cycle. For example, biologists investigate the life cycles of microbes, plants and animals and report their findings, usually in language that is obscure to the public. So to grasp what's missing in our conception of the dynamics of life, it sometimes helps to rely on the power of abstract thinking that typifies, for example, literature. Franz Kafka identified this 'missing piece' in his powerful novella *Metamorphosis*. Indeed, he captured it in the title: the capacity to change or transform from one state of body form to another. This is actually what evolution represents. It is the constant changing of living forms that we can trace on the tree of life. While individual species strive to maintain the immutability of the body's form, evolution strives for the change of one body form to another – the endless growth of branches on the tree of life.

The main character of *Metamorphosis*, Gregor Samsa, finds himself in – this is an understatement – an unusual situation: 'One morning, as Gregor Samsa was waking up from anxious dreams, he discovered that in bed he had been changed into a monstrous verminous bug.'[14] While humans can transform mentally, the physical transformation of our bodies – from human anatomy to the anatomy of a 'verminous bug' – is not possible. The efforts of Gregor Samsa's body to transform back are doomed to failure. His body is punished with a lack of ease until he dies. The novella's message is that the process of body transformation as a way of overcoming lack of ease is unattainable for the human species. The bodies of most multicellular organisms are finished products. The prospect of dramatic body change is limited to those species that can metamorphose: insects, fish, molluscs, amphibians, etc.[15]

Futurists and transhumanists may object to the assertion that the human body is a finished product. Ageing, according to some biologists, is a reversible process. Medical interventions, so the futuristic story goes, may turn the human body from its vulnerable state, prone to diseases and ageing, to a 'dis-ease'-free state of extended youth.[16] Humans will, in a not-so-distant future, live for an entire millennium. An even more optimistic prediction is that medicine will eradicate death.[17]

However, the transformative potential of life that keeps evolution moving, and the tree of life growing, is separated from the bodily structure, or soma, of animals and plants. That potential – the open evolutionary experiment that manifests in our imagination as the tree of life – is confined to germline cells. Our bodies are composed of trillions of cells we call somatic cells. These cells join forces and turn into a collective of cells, or the corporate body, in which the cells divide labour to produce tissues and organs. The corporate cellular collective descends from a single hybrid cell, the zygote, produced by the fusion of two germline cells: an egg and a spermatozoon. The transformative evolutionary potential, in the case of multicellular organisms, requires the fusion of two individuals, or more precisely the genetic essences of those individuals preserved in their germline cells. Germline cells, not corporate bodies, fuel evolution.

How can we link the evolutionarily transformative potential of germline cells with the concepts of lack of ease and ease of existence? Here we need to rely on microbes and the revolutionary work of Lynn Margulis. Bodies of bacteria are single cells, similar to the bodies of eggs and spermatozoa, but much less complex. As long as bacterial cells remain independent units, they function in the ease-of-existence mode. Bacterial cells contain minimal genomes and a minimal set of protein-mediated metabolic reactions, inside the boundary of the cell wall. This minimal biological identity allows bacterial ease of existence within the community of bacteria, including throughout the process of communication (see 'Kingdom Bacteria', page 90).

Lynn Margulis has shown us that ease of existence in the microbial world often turns into a nightmare – the lack of ease that occurs when two microbes (a bacterium and an archaeon) attempt to fuse.[18] Presumably, a larger microbe swallows a smaller one. Alternatively, a smaller microbial body infects a larger one. But the body of the smaller microbe resists the digestive apparatus of the host and attempts to retain its individuality. What happens next is a protracted struggle of two individual microbes to survive this monstrous change, to use the language of Kafka. Both microbes activate defence responses. Their genomes and cellular components enter a state of war – a lack of ease of epic proportions. The war is fatal for most attempts at microbial fusion. On rare occasions, however, the war ends in peace. The period of lack of ease turns into a period of ease of existence when the warring parties realise they can exist together in a new state. The new state is a form of biological novelty; formerly independent organisms merge to produce a new identity.

These microbial encounters, the monstrous fusion, the state of war and the occasional peace with which some encounters end is the process of endosymbiosis that Lynn Margulis described in her legendary 1967 paper and in subsequent papers and books. Thanks to microbes, which perfected the game of cellular fusion in deep evolutionary time by turning the initial monstrosity into a biological novelty, sex – a way of moving evolution forward – belongs to the category of ease of existence for us and all other animals.

In the act of sex, two germline cells fuse into one. This hybrid cell (the zygote) is a 'monstrosity' when you compare it with the parental germline cells. But the future of the zygote belongs to the category of ease of existence that we call development. The zygote will undergo embryogenesis, after which a new organism will be born. A new organism may belong to the same species from which the germline cells derive – that is, if the genomes of the germline cells remain within the genomic boundary of a species. However, if the genomes of the germline cells change sufficiently, a new organism may be born that differs from the species from which the germline cells are derived. The second possibility leads to the emergence of new species that enrich the tree of life.

Principles of Self-Preservation

We can now summarise life from the joint perspective of naturalised medicine and evolution. This will enable us to give naturalised medicine its voice, later in this section. The key thing to note is that life is not a struggle. It was a struggle in the early stages when Earth was turning from a dead planet into a living one, but once life established itself, it became a force that may be characterised as the 'joy of living' (see Chapter 4), or a true ease of existence (see Figure 8.1).

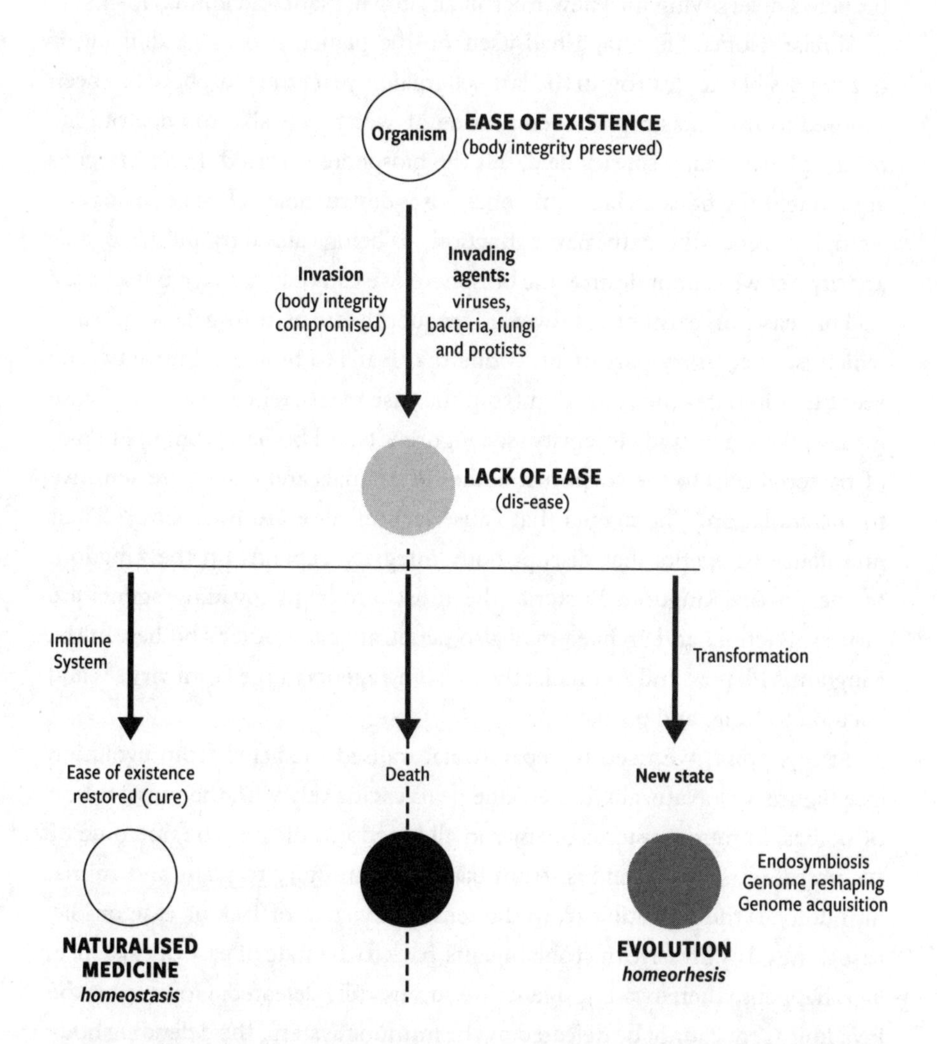

Figure 8.1. Evolution and medicine in the context of life as a planetary force. The 'organism' circle refers to all organisms, from bacteria to animals. The light grey circle is an infected organism. Infections have three outcomes: cure, death and initiation of bodily transformation. Endosymbiosis is a transformative event only in the case of prokaryotic microbes (bacteria and archaea). In all other organisms the transformative events range from genome reshaping to genome acquisition.

If life was a struggle, it would have probably remained a permanent monstrosity, always close to extinction and never able to spread across the planetary vastness. But thanks to the key feature of life – the ease of existence – life is a creative force. The tree of life is constantly growing, with branches diversifying into new microbial, fungal, plant and animal forms.

Similarly, once life established itself on the planet, it became difficult, if not impossible, to destroy. In the last 600 million years the biosphere has been exposed to five catastrophic events, none of which was able to wipe out life on the planet. Many species died, but the biosphere survived. Lynn Margulis argued that the basis of life – microbes – is indestructible. The last major catastrophic event – the sixth mass extinction – is being caused by anthropogenic activity. Yet we cannot destroy the biosphere; we can only damage ourselves.[19]

This ease of existence, however, frequently turns into a lack of ease, which is a necessary part of life – one that leads to biological novelty. The event that initiates the transition from the ease of existence to a lack of ease is the violation of body integrity (see Figure 8.1). All bodies, from the bodies of bacterial cells to the corporeal bodies of animals and plants, are sensitive to this violation. The agents that cause lack of ease are microbes.[20] Their prevalence as agents that disrupt body integrity depends on the kingdom of life. In the Kingdom Bacteria, the most dominant invading agents are viruses. Bacteria and archaea may also penetrate each other's bodies. In the Kingdoms Plantae and Animalia, the invading agents range from viruses and bacteria to fungi and protists.

At this point, we need to separate naturalised medicine from evolution (see Figure 8.1). Naturalised medicine deals exclusively with the preservation of bodies; immune systems operate in all kingdoms of life. The outcome of successful immune reactions, from bacterial immunity to plant and animal immunity, is the transition from the temporary state of lack of ease or 'dis-ease', caused by invasive microbial agents, back to the state of ease of existence. This happens when invading agents are successfully defeated. However, if the invading agent cannot be defeated by the immune system, the defending body will die. Therefore, naturalised medicine is a counter-evolutionary process. The inbuilt immune systems in all life forms aim to preserve the status quo (the evolutionary forms that existed before the violation of body integrity).

On the other hand, evolution as a creative process that generates new living forms must go beyond the self-preserving efforts of naturalised medicine. In the case of bacteria and archaea, the status of lack of ease must have another route open to them. That route involves a protracted effort on the part of bodies' self-preserving capacities – protection against viruses or protection from each other – that operate at the edge of death, leading to the emergence of a new stable state of the body, dramatically transformed relative to the previous state (see Figure 8.1).[21] The formal name of this process is endosymbiosis. Symbiosis research was dismissed by leading scientists only a few decades ago and blatantly characterised as pseudoscientific. However, thanks to the groundbreaking work of Lynn Margulis and her predecessors, endosymbiosis is recognised today as the undisputed generator of biological novelty, at least in the microbial world.

In the case of plants and animals, the transformation responsible for the emergence of biological novelty takes place in germline cells. How exactly does this process work? It depends on the school of thought to which scientists subscribe. Neo-Darwinists insist on the power of gene mutations – that is, random changes in single letters of the DNA alphabet known as point mutations, or losses or gains of a few letters. Germline cells containing novel mutations introduce these changes into the genetic pool of a species in the process of hybridisation with partner germline cells from the same species, also known as sex. Mutations accumulate to the point that the genomes and phenotypes they produce transform into new biological forms. These are new species that emerge as new branches on the tree of life, in a slow but steady process of natural selection.

However, Lynn Margulis and Dorion Sagan argued against neo-Darwinism and proposed that genome acquisition was the origin of species.[22] The generator of biological novelty is HGT. Acquisition of viral, bacterial and even protist genomes by germline cells reshapes plant and animal genomes and turns them into sources of new organismal phenotypes, or new species. Eukaryotic genomes are ever-changing mosaics in which neo-Darwinian mutations are only a minor source of variation. Genomic landscapes are endosymbiotic micro-ecologies in which integrated microbial DNA sequences sacrifice some of their original function but retain enough of it to contribute to the novel features of the emerging species.

The view that genomes are ever-changing mosaics of DNA sequences is shared by Eshel Ben-Jacob and James Shapiro.[23] Ben-Jacob argued that genomes are not passive repositories of DNA sequences but 'organs' of cells, constantly responding to environmental changes. The cell is a collection of molecular agents with a common goal and mutual dependence, manifesting as a unified system – the self – that exchanges energy and information with the environment. In this view, the genome is an agent of cellular memory. Shapiro also views the genome as an agent of cellular memory, and far more complex than the passive neo-Darwinian genetic repository. Genomes change actively, in a process Shapiro calls 'read-write genome', as part of constant cellular modifications and inscriptions.

In the most powerful rebuttal of neo-Darwinism to date, based on cytogenetic and genomic research in the last seventy years, Shapiro identified numerous methods used by eukaryotic cells to reshape their genomes and thus facilitate rapid macro-evolutionary changes.[24] These methods include the dispersal of genomic networks via transposable elements, as demonstrated by Barbara McClintock; the extensive production of RNA sequences that do not lead to proteins, but instead serve as regulatory genomic elements, thus challenging one of the neo-Darwinian concepts of 'junk DNA'; natural and artificial interspecies hybridisation via rapid activation of transposable elements and the subsequent rapid emergence of novel species; and rapid genome restructuring in cancer and germline cells, known as chromothripsis (chromosome shattering), chromoplexy (chromosome braiding) and chromoanasynthesis (chromosome build-up).

We can conclude by stating that the alliance between naturalised medicine and evolution is most clearly visible on the tree of life. Branches on the tree last because the self-preserving efforts of organismal bodies are successful. On the other hand, new branches on the tree grow because evolution is capable of turning self-preservation into the creative force that generates new species. In eukaryotes, new species arise by fusion of branches, genome acquisition, the processes of 'read-write genome' and the series of genome-reshaping methods identified by James Shapiro.

In other words, the tree of life is not only a tree but also a web.[25] This means that evolution and naturalised medicine are inextricably linked. Naturalised medicine is akin to homeostasis (see Figure 8.1), or a steady state that keeps the body preserved. For this reason, cells developed powerful DNA repair processes to keep genomes, and thus species phenotypes, intact, but also other phenotype-preserving processes including immunity. Evolution, on the other hand, is akin to homeorhesis – a change from one body form to another, or a steady flow of biogenic forms. That said, we must also interrogate the original definition of homeorhesis as a steady flow, in light of the rapid genome-reshaping processes identified by James Shapiro that lead to the sudden emergence of novel species.

Microbial Immunity

Scientists have identified six different antiviral defence systems in bacteria.[26] Some of these also operate in archaea. The most basic defence, innate immunity, involves the destruction of viral DNA or RNA by naturally occurring enzymes called restriction enzymes. The first such enzyme was isolated from the bacterium *Haemophilus influenzae* in 1970, and a total of 3,600 restriction enzymes are known today, 600 of which are commercially available. In fact, restriction enzymes have become workhorses of molecular biology whose value has been recognised by the Nobel Committee. The 1978 Nobel Prize was awarded to three scientists who discovered restriction enzymes and explained their theoretical basis.[27]

Restriction enzymes recognise sequences of six DNA letters and cleave the DNA into pieces. This action disables the viral genome by inducing its fragmentation, thus preventing infection. The process is a double-edged sword, however, because bacterial DNA can also be cleaved. To protect their own DNA, bacteria use a specific form of chemical modification. Six-letter recognition sites, or restriction sites, in the bacterial genome are modified by the addition of a chemical methyl group. This modification prevents restriction enzymes from recognising restriction sites present in the bacterial genome. The result is that only viral DNA is targeted and destroyed, while bacterial DNA remains intact.

Another bacterial antiviral defence system that targets viral DNA belongs to the category of adaptive immunity. It is called the CRISPR-Cas9 system and it was discovered and named by a Spanish microbiologist, Francisco Mojica, in collaboration with his colleagues.[28] (CRISPR means 'clustered regularly interspaced short palindromic repeats', and Cas9 [CRISPR-associated protein 9] is an enzyme that cuts viral DNA.) The process of bacterial adaptive immunity involves the intake of viral DNA into the bacterial genome in the form of short viral sequences called CRISPR spacers; these serve as a form of immune memory. When viruses invade bacteria, CRISPR spacers are converted into RNA sequences, which, through the process of complementarity, recognise viral DNA and guide the Cas9 enzyme to cleave the viral genome, thus destroying viral DNA and preventing infection. The sophistication and precision of the adaptive bacterial immune system inspired molecular biologists to develop the most sophisticated method for editing the human genome. The method, now known simply as CRISPR, took biomedical scientists by storm and earned the two scientists who invented it a Nobel Prize for Chemistry in 2020.[29]

The remaining four antiviral defence systems in bacteria include chemical defence, abortive infection, signalling systems and not yet fully understood systems.[30] The chemical defence system uses chemicals that incorporate into viral DNA and prevent viral replication. The abortive infection system utilises chemicals that interfere with the viral life cycle by stimulating the lysis (disintegration) of viruses before they can assemble into infecting particles. An example of a signalling system is the combination of a chemical that senses viral infections and in turn stimulates another chemical that induces bacterial death. The induction of bacterial death prevents viral replication and thus protects the bacterial community by sacrificing the affected member. The last system, consisting of a series of processes, involves a range of chemicals whose actions are not yet fully understood. These chemicals are named after mythological deities such as Zorya ('dawn' in English), a deity in Slavic folklore associated with health; Gabija, the spirit of fire in Lithuanian mythology, which protects home and family; Shedu, a winged creature from Mesopotamian mythology; and Thoeris, a goddess of childbirth and fertility in the ancient Egyptian religion.

The existence of at least six powerful and versatile antiviral defence systems that bacteria and archaea have been using for billions of years is a sign of their impressive biological sophistication. Yet we easily dismiss microbes as primitive forms of life, lagging far behind human evolutionary achievements.[31] For example, when Emmanuelle Charpentier and Jennifer A. Doudna won the Nobel Prize in Chemistry 'for the development of a method for genome editing' based on the bacterial adaptive immune system CRISPR-Cas9, the announcement was greeted with incredible enthusiasm by many.[32] After all, CRISPR-Cas9 has the potential to cure human genetic diseases, and if that's the case, it could lead to a medical revolution and ease the suffering of many. Imagine genetic ailments such as cystic fibrosis, Fanconi anaemia or

Huntington's disease being cured in our lifetimes. So much immense suffering would come to an end. This is why we need medicine.

Yet the Nobel Committee's decision also reveals a deep-seated prejudice. The new technique would not be available to us if bacteria hadn't perfected it billions of years ago. When Mojica first attempted to publish his discovery almost two decades ago, his experience was similar to the one Lynn Margulis had several decades earlier – that is, Mojica's paper was rejected by several leading scientific journals, including *Nature*. In the end, he published this important discovery in a journal that was not even in the top ten journals of his choice.[33]

Today we know that while the discovery of a novel microbial immune system was received as a minor scientific contribution by a publicity-shy Mojica, it would also turn out to be a big step for 'big science' – that is, science dominated by money – for which the modest Mojica would receive little credit. Biomedical scientists, aware of potential financial rewards and working on the grounds established by Mojica, manipulated the bacterial CRISPR-Cas9 immune system and turned it into a powerful method for editing the human genome. Indeed, two groups of scientists (one of which includes Emmanuelle Charpentier and Jennifer A. Doudna) are still involved in a patent battle for control of the CRISPR industry that has so far generated $1 billion.[34] That the Nobel Committee so blatantly ignored Mojica's monumental contribution means one thing: big science is about big money and big medicine, and is utterly decontextualised from the much bigger picture of naturalised science and naturalised medicine.

Homo sapiens is the pale blue dot of the biological world, lost in the evolutionary grandeur. Yet this does not stop us from expressing our immense creative potential in the form of, amongst other things, novel medical treatments such as the CRISPR genome-editing method. But we also need to acknowledge biological reality if we truly respect science. Respect for science would have looked like awarding one-third of the Nobel Prize to Francisco Mojica. Emmanuelle Charpentier and Jennifer A. Doudna would not have been able to turn CRISPR-Cas9 into a genome-editing tool without Mojica's groundbreaking discovery. The Nobel Prize in Chemistry for the development of the CRISPR editing method was an opportunity for the scientific community to demonstrate that science offers a mature, responsible and objective look at the world. The chance was, regrettably, squandered.

Plant and Animal Immunity

We know that the biosphere existed as the microbial planetary cloud, attached firmly to the planetary surface, for almost two billion years – from the moment bacteria overwhelmed the planet as the bacteriosphere roughly three billion years ago, to the point when eukaryotic multicellularity emerged roughly one billion years ago. If the ancient biosphere resembled its current form, we can assume (for illustrative purposes only) that the thickness of the ancient microbial cloud was 70 kilometres and the cloud roots penetrated 20 kilometres into the ground. Viruses and bacteria were distributed all over the cloud, but not homogeneously. For example, only certain types of bacteria and viruses were able to inhabit the upper parts of the atmosphere, or several kilometres below the planetary surface. Also, the density of microbes at those extreme locations was lower than at the ground and in the oceans. Archaea, microbial fungi and protists lived only in certain parts of the cloud, mostly associated with the planetary ground and oceans.

Roughly one billion years ago, eukaryotic multicellularity emerged. The microbial cloud became enriched with macrobes – that is, plants and animals.[35] Ever since then, large bodies of macrobes occupying the planetary surface and the oceans have lived continuously in close contact with trillions of microbes. To survive this type of existence, plants and animals developed sophisticated defence strategies to fight viral, bacterial and fungal infections. Protists are less frequent infecting agents, while archaea are not known to cause human diseases, even though they possess the required molecular means.

Our understanding of plant defence strategies against microbes is biased to a certain degree because plants represent an important resource for human agriculture. Almost all known epidemics and pandemics in plants are identified in the agricultural context and are not representative of a wider picture of microbial–plant relationships. Examples of viral epidemics and pandemics identified by botanists include: peach yellows, phony peach, sugarcane mosaic, pear decline, the swollen shoot of cacao, plum pox, sugar belt yellows, tobacco veinal necrosis, etc. Examples of fungal epidemics and pandemics include late blate of potato, blue mould of tobacco, hop downy mildew, Dutch elm disease, St Anthony's fire, cereal rusts, white pine blister rusts, brown spot of rice, red root of sugarcane, etc.[36]

Investigating the relationships between plants and microbes in the context of forest biomes, where the human influence is non-existent, can rectify this bias, especially when we take evolutionary-scale relationships into account. The emerging picture is that mutualistic symbiosis between plants on one side and fungi and bacteria on the other is the driving force behind biomes' ecological resilience (see Chapter 6). This mutualistic symbiosis, or ease of existence, in a true sense started 400 million years ago in the case of fungi and 60 million years ago in the case of bacteria. The ease of existence of mutualistic symbiosis, however, must have been preceded by the struggle between plants and microbes, or a lack of ease in their relationships, for an unspecified period until the stable state was achieved.

By examining the period of struggle in the plant–microbe relationship, we can begin to understand the plant immune system. Scientists divided this period of struggle into four phases.[37] In phase 1, trees, through their roots, release photosynthetic products of their own making, including sugars, amino acids and some other metabolites, which attract soil microbes to the roots. One particular class of metabolites, strigolactones, attract arbuscular mycorrhizal fungi, which start invading root cells (see 'Construction of Forest Infrastructure', page 118).

In phase 2, trees initiate an immune response against the fungi. Botanists divide plant immune responses into two branches. The first branch includes recognition of microbes, or cellular damage induced by microbes, by a variety of cellular receptors present on the surface of plant cells. These receptors trigger a range of immune responses collectively known as pattern-triggered immunity. The second branch includes recognition of virulence factors induced by microbes inside plant cells by resistance proteins. These proteins activate a series of molecular signalling events, the aim of which is to suppress microbial attacks.

In phase 3, the fungi initiate suppression of plant immunity and start recruiting bacteria to form a new microbial community known as mycorrhizosphere bacteria. This phase is a real struggle between defence responses launched by the plant cells and counter-responses triggered by the fungi. In the final phase, or phase 4, a new relationship between plant and fungi emerges known as mycorrhiza-induced resistance or induced systemic resistance. Fungal communities, which now live inside the root cells, protect the trees from many harmful microbes residing in the soil.

The final phase heralds the return to the state of ease of existence. This is a true state of contented symbiosis. Hosts, via their roots, provide shelter and photosynthetic nutrients for friendly microbes. Friendly microbes, in return, provide minerals and some rare amino acids to their hosts, and also synthesise toxins that kill invading pathogenic microbes, protecting hosts from infectious diseases. Yet the investigation of plant immunity by many botanists is predicated on the concept of hostility between plants and microbes, while mutualistic symbiosis, as a far more effective anti-infection strategy, tends to be ignored.

Similarly, animal immunity is often reified as the encounter between two hostile forces: animal multicellular bodies and microbes. The consequence of this mechanised neo-Darwinian view is a failure to appreciate that all animals, including humans, exist and have coevolved within a sea of symbiotic microbes. It has been estimated that the human microbiome contains 2,000 species of bacteria.[38] Microbiomes are our permanent biological extensions that provide us with many metabolic benefits, from the microbiome–brain axis to healthy ageing. On the other hand, there are only 100 species of pathogenic bacteria, which we encounter only rarely and can resist effectively provided we – and our commensal bacteria – are healthy.[39]

In light of the newly discovered biological reality – that animals coevolve with microbial partners – animal physiologist Margaret McFall-Ngai suggested a radically different interpretation of animal immunity.[40] She argues that immunity is a consequence of symbiosis management. Invertebrates manage symbiosis with their microbial partners through innate or general immunity, which lacks the memory of previous encounters. This reliance on innate immunity reflects the composition of the invertebrate microbiota. Invertebrates have small numbers of resident bacterial species. As a result, their innate immunity treats all microbes indiscriminately – as unwelcome invaders.

On the other hand, vertebrates have far greater numbers of resident bacterial species in our microbiomes. As a result, vertebrates face the conundrum of distinguishing between large numbers of friendly and hostile microbes. This requires more sophisticated immunity – adaptive or specialist immunity that relies on memorising previous encounters with microbes through cellular immunity. Thus, adaptive immunity provides an appropriately versatile management strategy of symbiosis to achieve the distinction between friendly and hostile microbes, which cannot be achieved by innate immunity.[41]

Animal immunity also involves endosymbiotic strategies. This happens in the case of invasion by viruses, which often integrate into animal genomes – a process known as viral endogenization. For example, one particular family of viruses, retroviruses, integrates its RNA genome into animal genomes. Using an enzyme called reverse transcriptase, retroviruses convert RNA into DNA, which becomes part of the hosts' genomes. It has been estimated that the human genome contains 450,000 sequences that represent integrated retroviruses, or in the specialist jargon, 'endogenous retroviruses'. Integrated sequences of retroviruses constitute at least 8% of the human genome.[42]

These integrated viral sequences may convey immunity against exogenous viruses. For example, koalas in Australia were infected by a retrovirus that integrated into the genome of their germline cells. After several generations, koalas acquired resistance to the same retrovirus through a process called superinfection exclusion. Biologists speculate that the integration of HIV into human germline cells may lead to HIV resistance in a few generations.[43]

In summary, all kingdoms of life have highly specialised immune strategies that make naturalised medicine, or self-preservation, a counter-evolutionary process. This is not to say that naturalised medicine is negative in the context of evolution; it is the opposite. Without self-preserving capacities, all species would be short-lived and evolution would be endlessly chaotic. Naturalised medicine and evolution are different sides of the same biological coin.

From Animal Self-Medication to Human Medicine

A new evolutionary phenomenon emerged, with animals representing a distinct precursor of human medicine. Some animals turned into 'doctors' through the practice of self-medication: the capacity to use plants and non-nutritional substances to treat diseases. They also turned into 'pharmacists', a practice known in scientific circles as zoopharmacognosy.[44] The terms 'doctors' and 'pharmacists', in the context of animal evolution, should be treated with caution. As in the case of the term 'scientist' (see Chapter 7), I'm using 'doctor' and 'pharmacist' in a limited and metaphoric sense. The aim is to trace the evolutionary path from animal self-medication to the practice of human medicine as we know it today.

Let's start with animals that are distant from us in terms of evolutionary heritage, but very close to us in terms of social organisation: insects. Given that social insects live in tightly packed nests, they are under constant threat from microbial pathogens and epidemics. It has been known since the 18th century that workers of some ant species collect pieces of solidified conifer resin and place them in their nests.[45] Scientists discovered that this behaviour represents a form of prophylactic medicine. The resin contains chemicals with antibacterial and antifungal properties. The use of resin by ants, bees and wasps protects these insect colonies and their broods against bacterial and fungal infections.[46] Pieces of resin are identified and collected by insects most likely through olfactory cues.[47]

But insects are not doctors who pursue only one line of expertise, like the simple collection of antimicrobial materials. Similar to human doctors, wood ants combine antimicrobial agents to induce more potent killing of pathogenic microbes. The bodies of wood ants have numerous glands that produce large quantities of chemicals, including formic and succinic acids. These chemicals are sprayed onto enemies and prey, and are also used to disinfect nests. Scientists have discovered even more complex medical practices. Wood ants combine formic and succinic acids with resin pieces. Analysis has revealed that application of these acids to pieces of resin produces more potent killing of the fungal pathogen *Metarhizium brunneum* relative to the use of the resin alone.[48]

The medical talent of ants doesn't end with prophylactic medicine. Amazingly, certain ant species show a talent for war surgery. Some of us may be familiar with war surgeons as masters of black humour, from the legendary movie *M*A*S*H*, in which two brilliant surgeons, Hawkeye Pierce and Trapper John McIntyre, disobey the army's rules and play a series of pranks on their colleagues. Perhaps this is the human way of dealing with the cruelty of war. But some ant species are born warriors and are deadly serious about war surgery. Here is why.

Ants from the species *Megaponera analis* specialise in raiding termite nests.[49] These warrior ants start marching in columns upon receiving the signal from scouts. There is a division of labour in columns. Larger ants, or majors, break the defensive barriers built by the termites, while smaller ants, or minors, quickly go through the openings created by the majors to kill the termites and pull them out of the nest. After the raids are finished, orderly columns form again, in which majors carry dead termites back to the ant nest. Scientists estimate that such raids occur two to four times a day.

The job of warrior ants carries certain health risks. Entomologist Erik Thomas Frank noticed that some majors carry not termites but fellow ants back to the nest.[50] Upon closer inspection, he noticed that all the ants being carried by fellow ants were severely injured. Some of them lacked legs; others had termites clinging from their bodies, the result of self-defensive bites. Frank devised a series of experiments to probe how the injured ants were treated. What he and his colleagues found was reminiscent of mobile army surgical hospitals. Injured ants were subjected to intensive medical care consisting of treatment of their wounds with antimicrobial chemicals, and surgery to remove clinging termites from their bodies. It is likely that *Megaponera analis* nests are full of mobile army surgical hospitals in which talented ant surgeons operate. The ant surgeons are so brilliant that discharged patients can be spotted on the battlefield the next day.

Self-medication has been observed not only in insects but in many other animals as well. Scientists use four criteria for detecting self-medication.[51] First, a plant used for self-medication must not be part of an animal's regular diet. Second, the plant should not provide any nutritional value to the animal. Third, the consumption of the medicinal plant must occur during the part of the year when infection frequency is high. Fourth, unaffected animals must not be involved in the practice of self-medication. Using these criteria, self-medication has been observed in a variety of animal species from different phyla, including bears, deer, elks, porcupines, jaguars, lizards, fruit flies, butterflies, elephants, woolly spider monkeys, lemurs, baboons, great apes and some domesticated animals such as goats and llamas.[52] The practice of self-medication in animals covers an evolutionary period of approximately 400 million years.

The practice of self-medication emerged in our hominin relatives, and in *Homo sapiens*, most likely through (1) sharing information between groups, including different hominin species, and (2) observing the actions of other animals. The evidence for sharing information between different hominin groups is solid. For example, archaeological records show that different groups of palaeolithic hominins practised self-medication a few million years ago using the same plants.[53] Similarly, medicinal plants and fungi that are still in use by modern humans, including yarrow, camomile and the fungus *Penicillium rubens* (the source of penicillin), were also used by Neanderthals 50,000 years ago.[54]

The evidence for observing the medicinal practices of other animals and learning how to treat humans comes from studies in ethnomedicine, ethnoveterinary medicine and ethnopharmacology. Michael A. Huffman, an expert in traditional medicine and animal behaviour, searches for medicinal plants discovered by animals. Huffman collaborated with the grandson of the medicine man Babu Kalunde of Tanzania, Mohamedi Seif Kalunde. The story, presented by Huffman, reveals how traditional medicine saved the lives of many people in Kalunde's village who were suffering from a dysentery-like illness.[55] More than a century ago Babu Kalunde, being an avid observer of wild animals, spotted a sick porcupine with symptoms of blood in its stool. He also noticed that the sick animal dug and chewed the roots of the plant *Aeschynomene cristata*, known by the locals as a poisonous plant called *mulengelele*. Further monitoring by Kalunde revealed that the porcupine fully recovered from the illness. Babu Kalunde tested the plant on himself and also persuaded the villagers to take *mulengelele* when they had dysentery-like symptoms. Kalunde hit the medical jackpot. The *mulengelele* root likely contains natural antibiotic substances, and Kalunde's grandson, Mohamedi Seif Kalunde, was able to expand the treatment to sexually transmitted diseases, such as gonorrhoea and syphilis.[56]

Another example includes collaboration between pastoralists and healers in the Karamoja region of Uganda to observe the self-medicating practices of livestock and apply some of these practices for human medication.[57] Readers interested in animal self-medicative wisdom can find further examples in the works of Michael A. Huffman.[58]

The 400-million-year history of self-medicating practices in animals, together with the traditional knowledge of hominins, Neanderthals and indigenous *Homo sapiens* societies, represent the evolutionary continuity of naturalised medicine (see Figure 8.1). Taking all the above evidence together, it seems appropriate to search for the roots of modern medicine at the interface between the prehistorical heritage of traditional knowledge and the more sophisticated philosophical knowledge that emerged with the Ancient Greeks. For example, Aristotle wrote about animal self-medication in Book VIII of *History of Animals*. It is unlikely that he personally observed animal self-medication cases. Instead, he relied on eyewitness accounts, some of which may have been influenced by Egyptian beliefs known to have been strongly held by the Greeks.[59] Given that Indian and Chinese medical texts existed at the time of Egyptian culture, it is likely that there was some interchange of medical practices between different cultures.

We must not forget that the transmission of knowledge was affected by the social organisation of various societies. Palaeolithic hominins, Neanderthals and indigenous *Homo sapiens* were hunter-gatherer societies. Their belief systems, including medicinal knowledge, survive until the present day. The same belief systems, or fragments of them, must have been preserved and promoted further, in the newly emerging human agricultural societies that sprang up at several different locations simultaneously from South America to the Middle East roughly ten millennia ago. The change in lifestyle from hunter-gatherer societies to agricultural societies facilitated the development of early civilisations, including the Babylonian, Mesopotamian, Egyptian, Indian and Chinese cultures, all of which practised some form of medicine. This rich multicultural inheritance formed the basis from which modern medicine emerged, starting with Hippocrates and Gallen in Graeco-Roman cultures, continuing with Islamic medicine and European medieval medicine, and culminating with the Enlightenment and the introduction of science into medicine. (We must not forget traditional Chinese medicine, which is not based on Western science.)

One of the advances of science-based medicine was the germ theory of disease, advocated by giants of modern medicine Ignaz Semmelweis, Louis Pasteur and Robert Koch. The germ theory of disease nicely fits with the naturalised version of medicine presented in Figure 8.1.[60] The discovery of microscopy enabled Antonie van Leeuwenhoek to observe bacteria in the 17th century. Viruses were discovered at the end of the 19th century. So we humans learnt, only a century or so ago, what bacteria 'knew' for billions of years – body integrity is compromised by the invasion of strange biogenic creatures that are neither dead nor alive (viruses) and that cause lack of ease within the invaded body. With the accumulation of modern knowledge, from Greek philosophy to science, we expanded medicine into areas never seen in evolution before. Are we going to become the first species to stop evolution by preventing the lack of ease that works in favour of evolution? If you believe futurists, the answer is yes. Medicine is geared towards establishing members of *Homo sapiens* as immortal deities above evolution. But this also may be a delusion. The COVID-19 pandemic teaches us that no matter how advanced civilisation may become, we are destined to live with viruses and bacteria forever.

Homeostasis versus Homeorhesis

Although conventional academic wisdom recognises medicine as an exclusively human invention, a naturalised form of medicine – the protection of body integrity from microbial invasion – is in fact widespread in nature (see Figure 8.1). All organisms, from bacteria to hominins, have self-preserving capacities in the form of various immune systems. Only in the Kingdom Bacteria does immunity preserve single cells. In all eukaryotic organisms, from protists to animals, immunity appears to represent various forms of symbiosis management, from the mutualistic symbiosis between plants and fungi to the adaptive immunity in animals that is responsible for distinguishing between friendly and pathogenic microbes. With the emergence of animals, self-preserving capacities entered another dimension. Some animal species use plants and other natural products to self-medicate. The roots of human medicine are in the 400-million-year history of animal self-medication.

I would like to go a step further and suggest that naturalised medicine is an integral part of evolution. Without the self-preserving capacities of organisms – naturalised medicine – evolution would be chaotic. Species would last only for a few generations and the tree of life would not be a tree at all, but a morass of short-lived biogenic monsters. Naturalised medicine, in the form of various types of immunity and self-medicating practices, gives stability to biological forms and enables them to last long enough to take their place on the tree of life.

Naturalised medicine (the self-preservation of body forms) and evolution (the transformative potential of one body form into another) are two sides of the same coin. To formalise these sides, let's call the self-preserving side homeostasis – the maintenance of the bodily steady state, specific for each species and let's call the transformative side homeorhesis – a flow of body forms that is not always even, but may be punctuated by more rapid flows (see Figure 8.1). One side complements the other.

CHAPTER NINE

Artists

I propose that the disciplines of aesthetics, art criticism, and art history should encompass both humans and non-human organisms, and that they should span evolutionary biology, behavioral biology, psychology, and the humanities.

RICHARD O. PRUM[1]

While it may seem strange to envisage non-human organisms as scientists, doctors, communicators or engineers, it is not at all strange to envisage them as artists. The first person who alluded to the animal artistic sense was none other than the greatest naturalist of all time, Charles Darwin. In *The Descent of Man, and Selection in Relation to Sex*, Darwin argued that the aesthetic mate choice drives evolution independent of natural selection.[2] Females of many bird and fish species select male partners by assessing their artistic sensibilities – from ornamental displays to the construction of love nests. Producing and evaluating beauty is an integral part of the lived experience that leads to procreation. Aesthetics shape evolution, according to Darwin.

Other leading scientists during Darwin's time didn't like this idea.[3] Alfred Russel Wallace was a rigid defender of natural selection as the only driver of evolution and dismissed the power of the aesthetic mate choice. His opinion resonated more with prevailing scientific opinions of the time than did Darwin's more 'dangerous' idea.[4]

But dangerous ideas often outlive their opposition. Richard O. Prum is an ornithologist, the curator of vertebrate zoology at the Peabody Museum of Natural History at Yale University and author of the popular 2017 book *The Evolution of Beauty*. Prum developed a theory of art that extends beyond conventional understanding. According to Prum, art is not confined to the activities and interests of humans. Instead, it is a biotic phenomenon; many non-human organisms possess artistic agency. Art is a population game in which the producers of art, usually males, are subjected to acts of evaluation by critics, usually females.[5] This artistic game between producers and critics ends in procreation. The product of the game – the offspring generated through the aesthetic mate choice – is not consistent with the concept of fitness promoted by neo-Darwinian biology. The fittest organisms, and their artistically most aware counterparts, are usually at different ends of the functional spectra in many species. Richard O. Prum revived Darwin's dangerous idea and took it to a new level.

A weakness of Prum's theory, however, is that he did not connect it to other naturalist theories of art. For example, the American philosopher John Dewey developed a powerful theory of art that viewed aesthetics as part of lived experience: the art of doing.[6] That is, any 'live creature' can have artistic experience through interactions with its environment[7]: 'For only when an organism shares in ordered relations of its environment does it secure the stability essential to living. And when the participation comes after a phase of disruption and conflict, it bears within itself the germs of a consummation akin to esthetic.'[8]

In the context of biocivilisations, aesthetics is an integral part of all living processes. It can be observed in many ways and in many places, including within the context of bacterial communication, the ways in which plants manage habitats and how animals select their mates. We also see it in the context of how human beings practise science and Gaia's hyperthought.

Producers and Evaluators

In 1995, a group of divers spotted unusual circles roughly 2 metres in diameter on the seabed just off the Japanese island of Amami Oshima. The complex and beautifully patterned circles lasted until the currents erased them. But even as more divers observed the circles, no one was able to explain their origin. This unusual phenomenon was labelled 'mysterious circles', and even 'underwater crop circles'.[9]

Information about the mysterious circles reached Hiroshi Kawase, the curator of the Coastal Branch of the Natural History Museum at Chiba. Kawase joined forces with marine biologists Yoji Okata and Kimiko Ito to investigate the origin of the circles. After several years, Kawase and his colleagues were able to identify what produced the circles: males from the pufferfish species *Torquigener (Tetraodontidae)*.[10] The species was later named *Torquigener albomaculosus*, or white-spotted pufferfish.

Pufferfish are well known to experts, but the novelty in the case of the white-spotted pufferfish was that nobody had ever seen these circles before. The small fish, only 12 centimetres long, had created 2-metre ornamental circles in the sand using its fins and body as a motor, taking about a week of relentless labour to complete these beautiful seabed creations.[11] Their purpose? Procreation. The producers of the ornamental circles are males; the evaluators of these artistic productions in the sand are females. If the females are sufficiently impressed by the beauty of the sand ornaments, they signal to the males their willingness to mate.

The story of the mysterious circles and the discovery of their creators was publicised widely. This caught the attention of the great British naturalist Sir David Attenborough and his team at the BBC. Attenborough and his team travelled to Japan to observe pufferfish males in action and produced a three-minute video called 'Courtship', which is part of episode 5 of BBC Earth's *Life Story*.[12] In his narration, Sir Attenborough went so far to describe the male pufferfish as 'nature's greatest artist'.

This compliment to the small fish, while lovely, is almost certainly an exaggeration, however. The male pufferfish is hardly unique in its artistic agency. Indeed, several years after the Japanese discovery, scientists spotted new circles 5,500 kilometres away, off the coast of Australia, at a depth of more than 130 metres (the Japanese circles were located at a depth of 30 metres).[13] The discovery of the Australian circles raised a lot of questions, particularly around the difference in depth. Was it possible that a different pufferfish species, unknown to science, was behind the Australian circles? Furthermore, it is dark in the ocean at a depth of 130 metres; how could the Australian females see the circles well enough to evaluate them in such dim conditions? If Japanese and Australian pufferfish are different species, they must have evolved such skills separately.

According to Prum, art is a nature-wide phenomenon, which he refers to as 'biotic art'.[14] This has profound implications. Do all species engage in acts of art? Is Gaia the biosystem influenced, or perhaps even dominated, by biotic art? For that matter, is naturalised science a *form* of art?

Even though Prum's theory cannot answer these questions, it offers a suitable platform from which to pose them, and the most important question is:

how far can we expand art and aesthetics from the narrow horizons of human culture to (possibly much wider) natural horizons? We already know that many non-human organisms possess artistic agency – the capacity to demonstrate autonomous preferences based on cognitive evaluation of sensory information.[15] Translated into plain language, artistic agency represents the capacity of an organism to produce, sense and evaluate beauty. White-spotted pufferfish, songbirds, ornamental birds, flowering plants, rattlesnakes, foraging insects and many other organisms are producers and evaluators of beauty.

Once we fully appreciate the essence of Prum's theory – the poverty of anthropic aesthetic parochialism – we can advance a step further and combine Prum's exploration of the subjective sense of beauty with Dewey's exploration of the objective sense of beauty. This peculiar combination represents a reasonably strong platform for answering the above questions in a meaningful way.

Art, according to Prum, represents a form of communication between the producers and evaluators of art.[16] Male pufferfish produce love nests in the sand and allow females to evaluate their merit as the entry ticket to the game of procreation. Likewise, flowering plants produce flowers of astonishing beauty and allow insect pollinators to evaluate them. The key here is communication. As we know from Chapter 5, the default form of communication in nature is biosemiotics. At the heart of biosemiotics is the triadic sign of object (O), representamen (R) and cognitive interpretation (I), ORI. Every biological sign carries within itself the abstract surrogate of the O it represents; the precise identity of that object, or its R, which makes it unique amongst numerous other objects; and finally, the possibility to link O and R through the process of I. The basic structure of the sign is the same for all organisms. There is no reason to believe that the human aesthetic experience, as a form of communication, is unique.

Interestingly, Prum does not use biosemiotics as an explanatory tool. This could be a simple case of ignorance because biosemiotics is a fringe discipline poorly known outside a small circle of experts. Yet Prum's unwitting reliance on biosemiotics is unmistakable:

> *In general, coevolved biotic signals in nature function explicitly as advertisements to other organisms. These signals include advertisements of sexual availability, desire (such as baby begging calls), the availability of ecological resources like nectar, pollen or fruit, and danger like the venomous*

> *rattlesnake's rattle or the bold markings of the noxious skunk. Biotic advertisements may function within a single species or amongst multiple species.*[17]

Let's use the relationship between flowering plants and insects to probe further. Both partners in this relationship evolve together – a clear case of coevolution recognised in Prum's terminology as 'coevolved biotic signals'. As the skills for recognising beauty grow in pollinating insects – their capacity to differentiate between a wide range of floral displays and remember those that carry the most attractive rewards – they stimulate the enhancement of visual and olfactory displays, or 'advertisements', in flowering plants. The coevolutionary relationship between flowering plants and insects is truly aesthetic; flowers that are pollinated by the wind lack aesthetic qualities.[18]

Here is a sketch of flowering plants' behaviour in the biosemiotic artistic context (i.e. in the context of reading and interpreting the triadic structure of the sign). In the cognitive space of a flowering plant – its *Umwelt* – the presence of flying beings that often land on flowers represents the O component of the sign that heralds the game of procreation. The act of pollination – the transfer of pollen grains from an antler to the stigma of a plant by insects – is the R that will materialise in the fertilisation and subsequent production of fruit. The third element of the sign, I, is the process of making sense of the sign within the cognitive space of the flowering plant.

In the *Umwelt* of insects such as honeybees, the O component of the sign is the physical appearance of flowers identified visually. The R component of the sign emerges when honeybees use olfactory and taste cues to evaluate the quality of nectar and pollen. The I element of the sign is the process of making sense of the sign within the cognitive space of honeybees.

The elaborate biosemiotic sign structure, which carries within itself the entire methodology required for biotic communication, incorporates Prum's definition of artistic agency. The cognitive evaluation and sensory information are already integrated into the process of semiosis via the O, R and I components of the sign. The same is true for biotic signals and advertisements. Therefore, the biosemiotic angle makes Prum's theory scientifically clearer because it replaces a somewhat arbitrary terminology (biotic signals, advertisements, etc.) with the clear terminology of biosemiotics (O, R, I).

The coevolutionary relationship between flowering plants and insects means that there must be an intensive sharing of semiotic spaces. This leads to reciprocal modifications of coevolving organisms. Insects constantly enhance their 'expertise' of the floral world through the I component of the sign, which involves cognitive evaluation. This has a reciprocal influence on the I component of the sign within the cognitive world of flowering plants. The growing expertise of insect evaluators makes the advertisements of floral displays more elaborate.

The sharing of semiotic spaces is called semiotic scaffolding – the network of semiotic interactions between organisms.[19] The sharing, in Prum's language, may be within a single species or amongst multiple species. If sharing occurs within a single species, it covers communication within a single cognitive world, as in the case of communication between male and female pufferfish. If sharing occurs between different species – flowering plants and insects, for example – it covers communication between two different cognitive worlds. However, even though flowering plants and insects are involved in the close coevolutionary relationship, they remain open to semiotic influences from other organisms. The extent of sharing semiotic spaces is so wide that biosemioticians interpret the biosphere as the semiosphere (see 'A Short History of Biosemiotics', page 86). Thus, in the context of Prum's theory, the semiosphere acquires elements of art and aesthetics. This new vision of the biological world directly challenges the concept of survival of the fittest: 'Nature in fact is not so much about "tooth and claw" as it is about sensing, interpreting, coordinating, and social co-operation.'[20]

Art of Doing

Prum's theory of art focuses on subjective aesthetic experiences. Communication channels between the producers and evaluators of art are biosemiotic channels between pairs of coevolving individuals. Dewey's theory of art, on the other hand, introduces a third party into the aesthetic experience – the environment – thereby objectivising aesthetic experience.[21]

The process of subjective aesthetic communication is so intense that the actors do not pay any attention to the outside world. The male pufferfish is concerned only with how his artistic circles in the sand will capture the attention of a female and entice her into the love game. He is not concerned at all with the sea bottom as his artistic platform, or other organisms that he encounters while ploughing the sand during this intense creative act.

Similarly, pufferfish females are only evaluating the beauty of the patterns in the sand, ignoring everything else. Two minds – the mind of the artist and the mind of the evaluator – are so intensely engaged in biotic communication that it is as if the external world does not exist.

Dewey exposes highly subjective aesthetic experiences to the judgement of the external world (in the context of the semiosphere, this means that communication channels between coevolving producers and evaluators of art become open to the flow of information from other parts of the semiosphere). Dewey's synonym for the external world is the environment.[22] Even though one can say that flowers are the environment for honeybees and vice versa, the objective sense of the environment is masked by the engagement of actors in coevolving semiotic spaces – so much so that all other communication channels with the external world are effectively closed.

When other communication channels are open, the actors get a wider sense of the external world and become engaged in other aesthetic processes. For example, honeybees use a surrogate of mathematics to precisely calculate the value of nectar from a given floral field and distribute workers to optimally exploit the field for nectar collection. People have observed that the relationship between honeybees and flowers is functionally and economically highly effective and have adopted it to solve a technological problem facing humanity. This is known as the honeybee algorithm.[23] An expert on honeybees, Thomas D. Seeley, worked with a team of engineers to copy the mathematical relationship between honeybees and flowers and applied it to the problem of internet traffic. The elegance of the solution and its financial value ($10 billion in savings for internet providers) prompted the prestigious American Association for Advancement of Science to award the Golden Goose Award to the researchers.[24]

The honeybee algorithm story aligns with Dewey's theory of art as the art of doing or making. The aesthetic emphasis is on the quality of the practical work carried out by honeybees for the benefit of themselves and flowers, which extends beyond their mutual relationship and into the cognitive world of a third party (human beings) who appreciate the honeybees' art of doing and have borrowed it to resolve their own practical problem.

Thus, Dewey sees art differently from Prum. While biotic art, according to Prum, is confined to aesthetic mate choices, ornamental advertisements and birds' songs, all of which represent subjective aesthetic experiences, art, for Dewey, is 'prefigured in the very process of living'.[25] That prefiguration is about the aesthetics of biological processes. Birds and beavers alter the environment by building nests and dams. For most people, these activities do not represent art, but for Dewey, these are examples of biotic art because birds' and beavers' activities are based on the interplay of natural energies from which the art of doing, or making, emerges.[26]

Certain metaphors reveal the aesthetics of biological processes, often brutally suppressed by the dry vocabulary of science. For example, the term 'photosynthesis' describes the process of converting CO_2 and water into energy, carbohydrates and oxygen by plants, with the help of sunlight. Biochemistry textbooks describe the process with cold precision, leaving no room within the description for anything that even remotely resembles art.[27] Yet the term 'eating the Sun' describes the process of photosynthesis with artistic flair and emphasises the art of doing.[28] Cyanobacteria learnt three billion years ago, how to make a feast out of almost nothing: a few micro-drops of water, mixed with air and spiced with a ray of sunlight. The cyanobacterial art of doing was so profound that it lives, through plants, until the present day, and enables us and other animals to exist.

Dewey, a naturalist par excellence, goes deep into the human world to explain the art of doing. The first thing he notes is that there is something wrong with our perception of art. If art is a natural phenomenon, humans have made the error of separating it from everyday experiences by confining artworks to museums.[29] There, artworks serve human vanity, rather than helping us understand life; this is akin to placing a songbird in a cage. The function of Dewey's theory is to correct the error of separating art from everyday experiences. Or, to remain within the metaphor, the function of Dewey's theory is to release the bird of art from the cage of museums.

The name that Dewey gives to the bird released from the cage is 'an experience'. This is a form of experience that has aesthetic qualities: 'A piece of work is finished in a way that is satisfactory; a problem receives its solution; a game is played through; a situation, whether that of eating a meal, playing a game of chess, carrying on a conversation, writing a book, or taking part in a political campaign, is so rounded out that its close is a consummation and not a cessation. Such an experience is a whole and carries with it its own individualizing quality and self-sufficiency. It is an experience.'[30]

However, an experience is a rare occurrence in human lives. We drift through life guided by the mechanics of the world we deliberately mechanised.

This drifting – the experiential process dominated by the succession of events that fail to unify into an experiential whole – immunises us against the true aesthetics of nature. Instead of striving to release the bird from the cage, we worship the cage. Dewey calls this 'anesthetic experience'.[31]

The dichotomy between an experience and anesthetic experience is powerfully expressed in the short poem 'Duck' by Vasko Popa[32]:

Duck
She waddles through the dust
In which no fish are smiling
Within her sides she carries
The restlessness of waters

Clumsy
She waddles slowly
The reeds she's thinking of
She'll reach them anyway

Never
Never will she be able
To walk
As she was able
To plough the mirrors

The duck is a metaphor for human nature; we are tortured by dichotomies of existence, including the struggle to distinguish between the art of doing and its opposite, the inability to appreciate that 'any practical activity will, provided that it is integrated and moves by its own urge to fulfillment, have esthetic quality'.[33]

The first two verses describe anesthetic experiences. Moving through the dust, heavily and awkwardly, is how the duck appears to us. The duck's fat, round body, equipped with exceptionally short legs, makes her movements inelegant and clumsy. Her world is dominated by the absence of the quality that could free her from the fate of the struggling and almost comic figure. That quality is expressed through the metaphor of the smiling fish. Smiling is a symbol of happiness, and the fish is a symbol of the water world. Indeed, the duck is better suited for the water world because her body can genuinely feel the fluidity or restlessness of water. Her short legs make her a comic figure on land, while in water her short legs become a perfect propelling and navigating instrument.

The tragedy of the duck's existence, in the wrong world, is further potentiated by the danger coming from 'a thinking reed' – a metaphor for humanity introduced by Blaise Pascal.[34] Because of her poor mobility on land, the duck is easy prey for humans. This seems to be her only purpose in the world.

Everything changes dramatically in the third verse, however. The surface of the water world is a mirror reflecting the land world. The duck's ability to slide through the water's surface transforms her from an inelegant, struggling land creature to a graceful and unique being that can 'plough the mirrors'. This is the true art of doing. Being able to plough the mirrors is the highest form of aesthetics in nature – the two worlds, land and water, are connected by a 'live creature'.[35] Fish are smiling. They are happy for the duck on behalf of everything that lives in the water.

Even though Dewey's theory of art is a naturalist theory, it has been rightly characterised as moderately naturalist.[36] However, replacing the terms 'an experience' and 'anesthetic experience' to describe human understanding of art in the naturalist context with poetic metaphors such as 'ploughing the mirrors' and 'waddling through the dust' opens the door for us to turn Dewey's theory into a strong naturalist theory. A strong naturalist theory of art encompasses the entire biosphere: biocivilisations become examples of the art of doing. As in the human world, dominated by streams of anesthetic experiences (waddling through the dust of everyday existence) and occasionally punctuated by episodes of an experience (ploughing the mirrors as a metaphor for the art of doing), the experiences of other organisms can be similarly understood. For example, pufferfish are artists only while the mating season lasts. At other times they are waddling through the dust of everyday existence.

An Indeterministic Gaia

These two naturalist theories of art complement each other. Prum's theory covers the subjective aesthetic experience of coevolving producers and evaluators of art. Dewey's theory objectivises the aesthetic experience by focusing on the art of doing, or the aesthetics of biological processes. Combined, the two theories of art allow us to go much deeper into the aesthetics of the biological world. We can now try to answer my earlier questions: do all species, and their biocivilisations, possess elements of biotic art? Is Gaia influenced, or perhaps even dominated, by biotic art? Is science a form of biotic art?

The first two questions are interconnected and can be reduced to one question because biocivilisations and Gaia are inextricably linked. Biocivilisations are the cognitive worlds, or *Umwelten*, of individual species, converted into biological substrates. Gaia is the totality of all the cognitive worlds and biological substrates that exist on Earth. These worlds are connected through biosemiosis. The totality and interconnectedness of cognitive worlds and substrates within the Gaian system are summarised in the term 'semiotic scaffolding'; the biosphere is the semiosphere. Gaia is the system that carries within herself her history; she 'feels' her own evolution. Gaia 'remembers' the past by maintaining the integrity of semiotic channels that go back to the origin of life. New organisms emerging in the evolution are immediately integrated into the network of ancient biotic channels. For example, we humans are starting to realise that we cannot exist without our microbial symbionts.[37] Even the development of our brains is influenced by our microbiome.[38] The veterans of evolution, bacteria, meet the evolutionary novice, *Homo sapiens*, and integrate us into the history of Gaia.

In the context of this book, hyperthought (see Chapter 2) is a synonym for semiotic scaffolding. Hyperthought is the Gaian mind; the mind that integrates, in a distributed and decentralised way, the actions of all the cognitive worlds in the biosphere – millions or even billions of *Umwelten*. This integration is a creative act that moves the Gaian system from one state to another, resulting in the emergence of new and unpredictable cognitive worlds and substrates (see Figure 2.1, page 44). Because the Gaian system cannot predict its next state, it is indeterministic, akin to a sculptor making herself without knowing her next move. The metaphor of Gaia the sculptor is the direct answer to the first two questions – now one. Gaia is not only influenced by biotic art but dominated by it. Before we go into detail about the aesthetics of Gaia, let's consider scientific arguments in favour of an indeterministic Gaia.

Stuart Kauffman argues that there are fundamental differences between physical and biological worlds.[39] While all possible stable atoms have been created in the long history of the abiotic universe, that is not the case with biomolecules such as proteins. The number of atoms in the observable universe has been estimated at 1,080. The number of all possible proteins with the length of 200 amino acids, on the other hand, is far greater: 10,260. It would take 1,039 lifetimes of the current universe to create all these proteins,[40] so existence above the level of atoms is non-ergodic.[41] This means that all possible life forms will never be created. Instead, biological entities, from cells to ecosystems to Gaia, exist as indeterministic holons – functionally closed biosystems that remain open to future evolutionary changes.[42] The consequence of non-ergodicity is that the biological world is part of the becoming of the universe – a process resistant to scientific predictability. The same may be true even of the abiotic universe. Attempts to formulate the Theory of Everything – the ultimate explanation of all aspects of the universe – are an epistemological error, according to Kauffman.[43]

Karl Popper argued similarly in his book *The Open Universe*.[44] If the universe were fully explicable by science – a possibility synonymous with the notion of a deterministic universe – we would be able to confidently predict all future events, under two conditions. We would need to have a precise description of all past events and we would need to know the laws of nature. Popper departs from the view that the universe is a deterministic machine (held by Spinoza, Hobbes, Hume, Kant and even Einstein) with a view of the universe as open, indeterministic and ultimately beyond the predictability of scientific theories.[45] Popper depicts scientific theories as a net designed by our rational minds to catch the fish of the world. He then argues that we will never be able to catch the fish because the net is always smaller than the fish (see Figure 9.1). Yet we should not give up on science because it is, along with art, a way for us to approach the open universe. As Popper said: 'The universe that harbours life is creative in best sense'.[46]

Recent work by Sergio Rubin and his colleagues shines new light on Kauffman's and Popper's contributions to our understanding of Gaia as an open and indeterministic system.[47] Rubin's main argument is that Gaia is an autopoietic and autonomous system. In light of Kauffman's view that biological systems are non-ergodic, and Popper's view that scientific theories cannot capture the real world, Rubin's arguments that autopoietic systems such as Gaia cannot be modelled by human scientific approaches are not surprising. This is the warning to scientists who think that humans can endlessly manipulate the biosphere without consequence and for self-serving purposes (see Chapters 3 and 7).

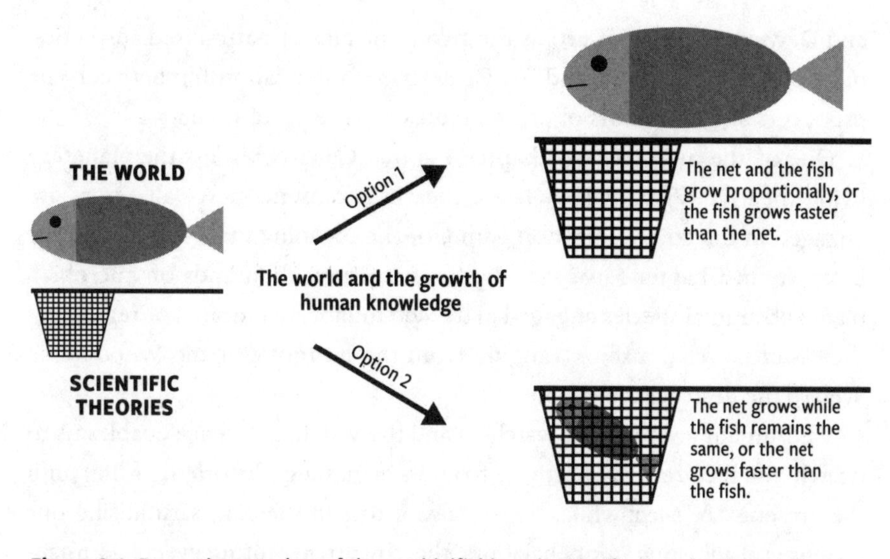

Figure 9.1. Popper's metaphor of the world (fish) and scientific theories (our rational net). Option 1 is more realistic than option 2. The net will always be smaller than the fish. Even if our net grows, the fish will grow proportionally, or faster.

Indeterminism, in the context of life, means that neither organisms nor the biosphere are mechanical, machine-like entities that can ever be fully modelled. Instead, life is characterised by a creative freedom that makes biological systems autopoietic and autonomous, open to future evolutionary changes. That is, biological systems are more complex than our models of them.[48] Hard-core scientists may counterargue that our scientific models, or the net for catching the fish, will one day outgrow the fish (see Figure 9.1). I disagree. As our net grows, so will the fish – even faster than our net.

The discrepancy between our rational net and the fish leads us into territory foreign to science. The biological world is aesthetic rather than mechanical. The implications of this view are profound. Nature is an open and creative system.

The Aesthetics of Gaia

What happens when we integrate naturalised communication (Chapter 5), engineering (Chapter 6), science (Chapter 7) and medicine (Chapter 8) with naturalised aesthetics?[49] To answer this question, I'll refer to Prum's and Dewey's theories of art, as the two principles of naturalised aesthetics, principles 1 and 2, or P1 and P2. P1 describes the relationship between the producers and evaluators of art; P2 represents the art of doing.

One of the messages of Chapter 5 is that Gaia represents the planetary living theatre. What makes Gaia a theatre is biosemiotics – all actors are engaged in the communication game, or the meaning-making process. For example, in Chapter 5 we saw the forest theatre: thousands of microbial, plant and animal species engaged in the 400-million-year drama of regulating their mutual relationships, ranging from the harmony of the Wood Wide Web to the destructions of war.

We humans are both the watchers and the watched. Science enables us to 'watch' the theatre retrospectively from its beginning (the origin of life) until the present day. Meanwhile, we are also actors in the play, scrutinising our actions and adjusting our behaviours, thereby driving future events. Humanity, as both the watchers and the watched within the living theatre, is a form of naturalised art. We produce and evaluate our own biotic art and evaluate biotic art produced by other species (P1). For example, traditional ecological knowledge suggests that certain ethnic groups, such as some North American Indigenous peoples, regard their actions in the context of the environment (for example, the use of land) as artful.[50] They evaluate their own artful actions by testing whether the use of land is sustainable, so that the land passed on to the next generation is of the same quality as that received from the preceding generation. Furthermore, some ethnic groups consider animals and plants as relatives and teachers. Respectful mimicry of animal behaviour represents the positive evaluation of biotic art. Likewise, plants may be considered fellow living beings rather than senseless, mechanical things.

Gaia, as a creative system par excellence, integrates all other creative forces and acts as the ultimate evaluator. But Gaia is also a sculptor making herself, autopoiesis being her technique of choice.[51] (I'm using the word 'technique' as Martin Heidegger did; *techne*, to the Ancient Greeks, meant both technique and art.[52]) Thus, the Gaian autopoietic technique is artful – a possibility directly in line with P2, the art of doing.

Natural engineering, in the context of an autopoietic Gaia – Gaia the self-sculptor – becomes biotic art. Examples of natural engineering include the construction of bacterial cities and forest infrastructure (see Chapter 6). Both of these examples of natural engineering are also examples of the art of doing (P2). Bacterial cities are metropolises with sophisticated internal structures consisting of the material secreted by bacterial bodies. Metropolises are linked with one another through biosemiotic signalling in the form of self-produced bursts of electrical activity. This interlinking enables metropolises to use food and other resources in an orderly fashion. Bacterial metropolises are building blocks of the bacteriosphere, the planetary microbial system.

Bacterial metropolises are also examples of the aesthetics of biological processes. The key biological process here is bacterial sociality. The bacterial metropolis is a form of biotic art that allows us to trace bacterial sociality from the single bacterial cell to the bacteriosphere, exemplifying Dewey's words that art is 'prefigured in the very process of living'. Bacterial biotic art is the basis for the life of plants and animals – the bacteriosphere has been regulating biogeochemical affairs on Earth for the last three billion years. So bacteria truly 'plough the mirrors' – they link the abiotic universe with the living universe.

The engineering of forest infrastructure also exemplifies the aesthetics of biological processes. This time aesthetics manifest in the art of symbiosis. Trees established a symbiosis with fungi and bacteria; this enabled them to trade nutrients produced through photosynthesis for the essential underground nutrients provided by fungi and bacteria. By controlling the symbiotic relationships with fungi and bacteria, trees acquired the capacity to produce biomes resilient to ecological changes, including shelter for numerous microbial, plant and animal species. 'Ploughing the mirrors', in this context, means that trees, through their roots, link the hidden underground world with its aboveground counterpart.

In Chapter 7, I argued that science can be naturalised by interpreting it as an autopoietic problem-solving process, applicable to all species. In the context of naturalised aesthetics, this means that biology can no longer be viewed in the same light as physics and inorganic chemistry. Planets, stones and atoms do not have subjective aesthetic experiences, while flowers, honeybees and many other organisms do. The creative freedom of biological systems makes them autopoietic and indeterministic. Ultimately, these systems are aesthetic ones that defy mechanisation. Dealing with aesthetic systems requires changes in the structure of science, however. The classical view, that biology is reducible to chemistry and physics, will no longer hold.[53]

To reach beyond the classical structure of science, we can combine biology with naturalised aesthetics. Once we do, biology becomes, in part, an exercise in evaluating biotic art. This also answers, in part, the question I posed earlier: is science a form of biotic art? In contrast to physics and inorganic chemistry, biology is a science that must take account of naturalised aesthetics. The term 'artful science' applies to biology. Fundamental discoveries in biology reveal the aesthetics of biological processes (P2). The power of biotic art is recognised by its originality and cannot be mimicked by human discoverers.

There are many examples of this from the history of biology, such as the discovery of restriction enzymes that protect bacteria from viruses (see 'Microbial Immunity', page 149). Restriction enzymes recognise precise sequences in viral DNA and cut them, destroying viruses while the bacterial DNA remains protected. It took many years of intensive work for scientists to identify the first restriction enzyme, produced by the bacterium *Haemophilus influenzae*, and to understand how it works. This cleared the way for the use of restriction enzymes as molecular scissors in genetic engineering and the science of genomics. The biggest international project in biology, the HGP, would not have been possible without restriction enzymes.

This is the art of doing. Bacteria's capacity to kill viruses by targeting viral DNA, while preserving their own, enabled the science of genomics, billions of years after the original bacterial 'discovery'. Bacteria 'plough the mirrors' by crossing the temporal and spatial borders of the kingdoms of life. The example also illustrates the hidden aesthetics of science. The most important discoveries are those that are often most elegant and far-reaching in their applications. For example, discoveries in microbiology extend beyond the peculiarities of bacteria and find applications in medicine. Acquiring values from non-human worlds is a feature of the human art of doing. This is also an unwitting recognition that biology is, in part, the evaluation of biotic art (P1).[54]

In Chapter 8, I contrasted naturalised medicine with evolution. While naturalised medicine is concerned with the self-preservation of body forms, or homeostasis (physiology), evolution allows changes from one body form into another, or homeorhesis (evolution). In the context of naturalised aesthetics, naturalised medicine becomes an example of the art of doing (P2). It ranges from bacterial immunity to the self-preservation efforts of animals (see Chapter 8). The art of doing in the bacterial world of naturalised medicine is so profound that today two groups of medical experts are fighting for the right to use the CRISPR-Cas9 method for medical purposes.[55]

Some forms of naturalised medicine are in line with P1 (the relationship between the producers and evaluators of art). Babu Kalunde, the medicine man from Tanzania, relied on animal expertise; he watched biotic art in action as he observed how the animals were using plants to cure or treat their ailments. Kalunde then tested the medical effect of the plants on himself (in a positive evaluation of biotic art) before introducing animal practices into traditional medicine.

Modern medicine, heavily dominated by the biomechanical models of molecular biology and genetics, has largely expelled artistic creativity from its educational and professional paradigm. Yet this does not stop some doctors from challenging the domination of biomechanical models. In a series of articles, a group of international doctors known as the Berne Group argued for supplementing biomechanical models with biosemiotic and autopoietic models to tackle the weaknesses of modern medicine, which is dominated by a 'monocausal, reductionistic view of health and disease'.[56] In other words, medicine is neither pure science nor pure art. As the 19th-century French clinician Armand Trousseau noted: 'The worst man of science is he who is never an artist, and the worst artist is he who is never a man of science.'[57]

Artful Biology

The abiotic world, from atoms to galaxies, appears deterministic and predictable. Some of the greatest human minds ever to grace planet Earth have put their faith in the ability of scientific theories to confidently retrodict and predict the universe. Black holes, gravitational waves, subatomic particles and other physical phenomena seem to obey mathematical formulas. The optimistic scenario is that one day we will have the final theory – the Theory of Everything – that faithfully describes the physical universe. Or will we? Stuart Kauffman thinks this is a delusion. Similarly, Karl Popper thought that the universe is 'partly causal, partly probabilistic, and partly open'.[58] The openness of the universe makes our scientific theories only partly effective.

The key feature of the universe, according to Popper, is creativity. This is most clearly visible in the biological world. The objects of our investigation are creative, autopoietic systems, from cells to Gaia. Scientists may think they can fully explain the biological world, but reductive materialism is limited when it comes to capturing the essence of life (see Chapter 2). The reductionist approach, according to Popper, may not hold up against the abiotic world either. We end up in a paradox of incompatibility between the fish (the world) and our rational net (scientific theories) (see Figure 9.1).

A resolution to this paradox is to acknowledge that biological systems (at least) are not mechanical, machine-like systems but autopoietic, indeterministic and creative ones. In order to have some understanding of these systems, we have to integrate science with naturalist theories of art, and rather than assuming that the biological world is here for us to catch in our rational net, we can attempt to re-envision it as a creative system that is willing to take us along on a journey into the unknown. But first we have to adjust our science of life to the challenges of the journey. The term 'artful biology' is not an oxymoron; it is an existential cry for a better epistemology.

CHAPTER TEN

Farmers

We simply took what was given us and continued to multiply and consume in blind obedience to instincts inherited from our humbler, more brutally constrained Paleolithic ancestors.

E.O. Wilson[1]

We have now explored biocivilisations from a variety of perspectives – arguably peaks of modern human civilisation – including communication, engineering, science, medicine and art (see Figure 5.1, page 81). The remaining aspect I've chosen to cover is agriculture, or population feeding. But as in previous chapters, we first need to make an imaginative leap over the anthropocentric walls so we can explore the true meaning of agriculture in the context of biocivilisations.

Consuming is a permanent feature of biocivilisations. In the case of animals, we call this feeding. Some animals spend their entire lives looking for a meal. The adult female tick will wait patiently after copulation until it detects a warm-blooded animal. She will then suck the blood of her prey as her last meal. Engorged with her prey's blood and unable to control her body, she will fall to the ground, lay eggs and die.[2] Luckily, meals are not life-threatening experiences for most other animals, but rather memorable episodes of tasting pleasures.

Given the importance of food for survival, the biosphere, in one of its numerous incarnations, is an endless feast. And the table manners at this feast are strange indeed. Guests can eat each other without any retaliation or revenge; this is predator–prey symbiosis. Indeed, one group of guests can completely devour the entire population of another. According to some scientific models, our hunter-gatherer ancestors exterminated entire populations of prey 13,000 years ago, inducing their own starvation.[3] This existential challenge may have prompted the invention of agriculture as a more effective way to feed large groups. And yet, as much as it might look like anarchy, the planetary feast is not a party for lunatics. There is a hidden order beneath it all.

Scientists Kenneth H. Nealson and Pamela G. Conrad have posited a view of life 'in terms of energy flows and metabolic capability', classifying organisms into three functional categories: physicists, chemists and biologists.[4] Physicists are organisms that use physical sources of energy, such as light; chemists use chemical energy, either inorganic or organic; and biologists feed on other organisms. Physicists and chemists are autotrophs; they are independent food producers, including microbes and plants. Biologists are heterotrophs – animals that eat other animals or plants.

This feast of life is a surreal phenomenon. Many important guests are invisible – bacteria, disguised as physicists and chemists, 'eat the Sun' and certain chemicals. Physicists and chemists create the invisible factory of essential food ingredients: biogeochemical cycles of carbon, sulphur and other organic elements that represent the building blocks of food. Some Sun-eaters re-emerge in a different guise, as endosymbiotic partners of plants, for example.[5] With the emergence of plants and animals, the feast became a frenzy of biologists gorging themselves in a paradise of food.

Yet this paradise is deceptive. Some organisms expend effort to produce their own food through agriculture. In the case of humans, we discovered agriculture roughly ten millennia ago and it became a springboard for economic surplus, a condition almost unknown to our hunter-gatherer ancestors.[6] This economic surplus facilitated discoveries including written language, mathematics, architecture, philosophy and science.[7] Many advanced human civilisations emerged in the last few millennia, culminating in modern civilisation, which exists at the edge of the cosmic adventure.

We are not the only farmers in the biosphere, however. For example, amoebas farm bacteria, and insects and snails farm fungi.[8] Agriculture, in the context of biocivilisations, is a form of mutualistic symbiosis beneficial for all partners. For example, aphid-farming ants protect aphids against parasites.[9] Ants, termites and beetles invented fungal farming forty to sixty million years ago.[10] Agriculture offers advantages to partners, including stable strategies for survival over millions of years.

On the other hand, human experience suggests that agriculture may also pose an existential threat in the form of uncontrolled growth and runaway consumption. Arguably, without agriculture, there would be no economic surplus. Without economic surplus, there would be no enlightenment or scientific advances, including the capacity to eradicate pathogenic microbes as the major exterminators of human populations.[11] The world population of humans – currently close to eight billion – consumes more planetary resources than any other species.[12] E.O. Wilson uses the abbreviation HIPPO to depict the devastating effects on the biosphere:

- Habitat loss
- Invasive species
- Pollution
- (Human) Population
- Overharvesting[13]

A modern-day Darwin, E.O. Wilson, criticised our folly: 'A few lifetimes from now, and then on for centuries to follow, humanity is going to be asking: what did we think we were doing?'[14]

Why do some farming practices (of insects, for example) last for millions of years without putting undue pressure on resources, whereas the relatively recent advent of human agriculture currently appears to be posing a rather immediate existential threat?

A Brief History of Agriculture

Scientists who investigate agriculture in the evolutionary context usually focus on social insects and humans. For example:

> *Insect fungiculture and human farming share the defining features of agriculture...: (a) habitual planting ('inoculation') of sessile (non-mobile) cultivars in particular habitats or on particular substrates, including the seeding of new gardens with crop propagules (seeds, cuttings, or inocula) that are selected by the farmers from mature ('ripe') gardens and transferred to novel gardens; (b) cultivation aimed at the improvement of growth conditions for the crop (e.g., manuring; regulation of temperature, moisture, or humidity), or protection of the crop against herbivores/fungivores, parasites, or diseases; (c) harvesting of*

the cultivar for food; and (d) obligate (in insects) or effectively obligate (in humans) nutritional dependency on the crop.[15]

The above description of agriculture is incomplete, however. To fully appreciate the history of agriculture in the evolutionary context, we have to identify missing elements. Let's start with the most obvious one: human farming of livestock, otherwise known as animal husbandry. Animal husbandry includes the production of meat, dairy, eggs, wool and other products, and may include the rearing of cows, sheep, goats, pigs and poultry.

There are also more obscure types of animal husbandry. For example, some animals are farmed only in certain places, including buffalos, alpacas and llamas. There are also farming practices that involve fish (aquaculture), insects such as honeybees (beekeeping) and small livestock such as rabbits and other rodents. Many people were unaware that mink are farmed for fur until concerns about cross-species infection during the COVID-19 pandemic led authorities to call for the culling of Denmark's entire mink population, a total of seventeen million animals.[16] Other unusual farming practices include sericulture, or the production of silk, which is produced by the glands of larvae of lepidopteran insects, as well as both juvenile and adult spiders. The commercial production of silk today largely relies on *Bombyx mori*, a domestic silk moth, also known as the Oriental Silkworm. Sometimes insects are farmed as food.[17] For example, selling and eating edible insects is a common practice in Thailand. With the human population predicted to exceed nine billion by 2050, the replacement of standard livestock with 'six-legged livestock' as an alternative to meat production is well underway.[18]

Animal husbandry can also take place outside human culture, such as the symbiotic relationship between ants and aphids.[19] Ants act as herders and protectors of aphids – sap-sucking, soft-bodied insects, also known as 'ant cows' or greenfly and usually no bigger than a pinhead. Aphids live on vascular plants, where they suck the nutrient-rich phloem sap from the plant tissue. Ants, in turn, 'milk' aphids to release honeydew by stroking the end of the aphid's alimentary tract using their antennae. The ants then transport the honeydew to their nests and regurgitate it for their nestmates. For some ant species, aphid honeydew is the only source of food. Occasionally, ants supplement their diet by eating aphids, most likely when the aphid herd becomes too large. In return for the food, the ants protect the aphids from predators and even provide medical care. For example, ants actively search for a fungus that infects aphids and remove it from the aphids' bodies.

The mutualistic relationship between ants and aphids is so strong that when young ant queens leave the nest to start a new colony, they will carry aphid eggs in their mouths to start a new herd. The strength of the mutualistic relationship is further illustrated by the farming skills developed by ants, and the confidence of aphids in their herders. For example, ants will secrete chemicals to drug aphids, thereby slowing their movements to prepare them for milking. For their part, the aphids will reduce their behavioural defence mechanisms.

Farming practices also occur outside the Kingdom Animalia. In Chapter 5, I covered the symbiosis between the social amoeba *Dictyostelium discoideum* and bacteria from the genus *Burkholderia*. Depending on nutritional conditions, *Dictyostelium discoideum* live either as solitary or social creatures. When the bacteria that serve as food for amoeba are plentiful, *Dictyostelium discoideum* live as solitary creatures, dividing by binary fission, but when bacteria are scarce, the amoebas will congregate to form a multicellular fruiting body, with around 20% of cells sacrificing themselves to its formation. The remaining cells ascend the stalk, called a sorus, to the top and form a globule that will differentiate into spores. Analysis of the spores has revealed that many of them carry bacteria. These spores are called 'farmers' because they can reseed the new source of food. Further analysis has revealed that bacteria are not passive partners in this farming symbiosis. Spores without bacteria ('non-farmers') will be colonised by *Burkholderia* to turn them into farmers.[20]

Finally, agriculture has also been discovered in the Kingdom Fungi. The soil fungus *Morchella crassipes* farms the bacterial species *Pseudomonas putida* by habitually planting the bacteria, feeding them on fungal exudates and harvesting them.[21]

In other words, agriculture is an evolutionary phenomenon much older than insect fungiculture, which is often interpreted as the sole precursor to human agriculture. The origin of agriculture can be traced to the prokaryote–eukaryote evolutionary transition. As soon as the first eukaryotes emerged, they started mixing with bacteria and discovered that they could eat them. In Chapter 5 we learnt that protists are expert readers of bacterial VOCs and can graze bacterial fields in the form of biofilms (preferentially). However, we have also seen that when entire local populations of bacteria are devoured, protists are forced, like social amoebas, to invent a new feeding technique: agriculture. The first farmers in the history of life were protists.

This revised history of agriculture, which pushes back its origin more than a billion years into the evolutionary past (from insect fungiculture to protists), has an extremely important implication. The key driver of agriculture is symbiosis. After the emergence of endosymbiosis as a major evolutionary transition roughly 1.5 billion years ago, symbiotic relationships between prokaryotic microbes and their eukaryotic counterparts exploded. The power of this symbiotic explosion catapulted agriculture, as a form of mutualistic symbiosis, far into the future. More than a billion years later, agriculture was discovered by social insects. Our hunter-gatherer ancestors unwittingly copied the actions of our agricultural predecessors in the long line of the evolutionary convergence that started with the protist–bacterial relationship.

Farming as a Social Game

Insect fungiculture remains the best-understood form of farming outside human agriculture. This is, in part, because insects and humans show similarities in social behaviour. The highest form of sociality in the animal world is known as eusociality. A group of animals is considered eusocial if three key patterns of social behaviour are observed: they care for their young, they have overlapping generations and they carry out division of labour, including reproductive labour (see 'Eusociality and the Origin of Cities', page 73). Some species of ants, termites and honeybees show the above behavioural patterns and are thus considered eusocial.[22]

There is no agreement amongst scientists, however, on whether *Homo sapiens* is a eusocial species.[23] The main obstacle to accepting human eusociality is the perceived lack of reproductive division of labour. In ant, termite and honeybee societies, which have either a queen (ants and honeybees) or a queen and a king (termites) as the only reproductively active members of the society, the rest of the society's members are sterile. But reproductive division of labour seems to be absent in human societies. For this reason, some scientists think that *Homo sapiens* is not a eusocial species.

Some scientists argue that the division of reproductive labour in human societies should be considered in the context of reproductive physiology, such as menopause – a programmed process of acquired sterility, also known as the grandmother effect. Raising children in humans lasts longer than in other animals, and a set of specialised skills is required that young mothers lack, so grandmothers become sterile carers of grandchildren. The skills their daughters learn will be passed on when the daughters become grandmothers. The post-menopausal 'caste' of grandmother helpers is the closest we get to the reproductive division of labour in human societies.[24] The grandmother effect prompted some scientists, including E.O. Wilson, to argue strongly in favour of human eusociality.[25]

What's the link between eusociality and agriculture? Even though agriculture is not a precondition for eusociality, the way biologists interpret eusociality mathematically reveals the importance of agriculture in some models of eusociality. Before I focus on mathematical models, it is important to outline the principles of ant agriculture in the context of their sociality. This will enable us to identify parallels between ant and human eusociality that are important for understanding the existential dangers behind human agriculture. For this purpose, we can return to leafcutter ants, one of the best-understood models of insect agriculture.

The agricultural symbiosis between leafcutter ants and fungi from the genus *Leucoagaricus* is about fifty million years old.[26] It happened for the first time in the area of today's Amazon rainforest and spread north all the way to New Jersey in the United States, resulting in many new ant species scattered from Argentina to the United States.

Leafcutter nests are engineering wonders, reminiscent of our cities (see 'Eusociality and the Origin of Cities', page 73). The 'buildings' in ant cities for the cultivation of fungi are called fungal gardens and they represent the digestive tract of the ant superorganism. Leafcutters are probably the most successful gardeners amongst insects. They bring pieces of cut leaves to the nest, which specialist workers chew and then deposit in the fungal gardens. Fungi are then seeded on the mulched green mass. They will produce nutritious hyphae called gongylidia, which feed the entire ant city.

Maintaining optimal conditions for fungal growth requires sophisticated agricultural skill: sowing the correct fungal mycelium on fresh soil, recognising and removing unwanted fungi, fertilising the substrate with their faeces, using antibiotics secreted by ant glands to prevent the growth of unwanted fungi, etc. A major danger (in all agriculture) is contamination by unwanted bacteria and fungi. The fungal parasite from the genus *Escovopsis* has survived for twenty-three million years thanks to leafcutter fungal gardens. Ants try to control the parasite by secreting an antibiotic fluid directly onto the spores of this fungus. However, the main weapon in the fight against the fungal parasite is a bacterium from the genus *Pseudonocardia*. This bacterium secretes a substance known as dentigerumycin, which acts selectively on *Escovopsis* and prevents its further growth. *Pseudonocardia* live on the bodies of leafcutters, so this agricultural symbiosis between ants and fungi cannot survive without bacteria.

The ecological success of eusociality is apparent in the production of biomass. Ants and termites represent only 2% of roughly 900,000 insect species, yet they constitute half of the insect biomass.[27] Interestingly, human biomass is equivalent to ant biomass. However, only 200 years ago our biomass was seven to eight times lower than it is at present. Our population growth appears to have coincided with the Industrial Revolution. Improvements in technologies from building skills and transportation to food production and medicine have resulted in the doubling of the human population in successively shorter time spans. For example, it took 260 years for the human population to grow from 0.5 to 1 billion, between 1534 and 1803. However, the growth from 2 to 4 billion occurred in only 47 years, from 1928 to 1975.[28]

It appears that the ecological success of human eusociality is linked with our technological development. In Chapter 6 we learnt that technologies are widespread in nature. If human eusociality is linked with technological development, insect eusociality may depend on insect technological developments. This thesis is supported by at least two lines of evidence.

First, analysis of ant agriculture has shown that it has all the elements of technology from Li-Hua's framework (see Figure 6.1, page 105).[29] For example, technique includes instruments and materials, and in the case of leafcutters, the instruments used for cutting leaves are sharp jaws, controlled by powerful muscles, that act as vibrating knives or microtomes. The transportation instruments are the six-legged bodies of ants that transport pieces of cut leaves from the cutting site to the nest. The material is the green leaf biomass and the symbiotic fungus. Knowledge is based on instinct. Leafcutters know how to cut leaves, transport leaves, establish fungal gardens, cure gardens of infections, etc. The organisation of work is inherent in the organisation of the ant society, which involves the division of labour. The product is the food for the entire ant colony.

Second, according to E.O. Wilson, a precondition for eusociality is the ability of social groups to establish defensible nests.[30] These defensible nests are ant and human cities. They share numerous features, including sophisticated buildings linked by complex transportation networks, side roads that lead to rubbish pits, army barracks, police stations, etc. (see 'Eusociality and the Origin of Cities', page 73). When scientists excavated an underground leafcutter nest in Brazil, they started by pouring 10 tonnes of concrete into holes at the surface of the nest. They discovered that the labyrinth of interconnected tunnels was so complex that it took ten days to pour the concrete down the network of tunnels and several months to excavate the entire nest. The solidified concrete preserved the tunnels and allowed scientists to investigate the underground city in astonishing detail. It covered a surface of 50 square metres, extended 8 metres below the surface and contained numerous chambers, or buildings, containing fungal gardens, which were, on average, the size of a human head. These garden buildings were connected by sophisticated highways through which worker ants transported pieces of cut leaves from cutting sites located dozens to hundreds of metres from the city. This analysis shows that, at least in the case of leafcutter ants, the engineering skills for city construction and the agricultural skills for the production of food on a large scale are inextricably linked. The defensible nest as a precondition for eusociality requires both engineering and agricultural skills.

Mathematics of Eusociality

We can now go back to the mathematical models of eusociality to probe the link between engineering and agriculture. The mathematics behind eusociality is heavily contested, involving two large disputes, the first in 1975 and the second in 2010, both of which involved E.O. Wilson. In 1975, Wilson published a seminal book, *Sociobiology*, which was attacked by a group of biologists, including Stephen Jay Gould and Richard Lewontin, in a letter to the *New York Times Review of Books*.[31] They accused Wilson of reviving the spirit of eugenics and even Nazism by interpreting human nature as the continuum of animal behaviour. In particular, they focused on Wilson's propensity to rely on social insect behaviour as the basis for understanding human sociality.

The mathematical model Wilson used in *Sociobiology* was developed by an Oxford biologist, William D. Hamilton, based on an idea first proposed by the British polymath J.B.S. Haldane. Hamilton's mathematical model paved the way for kin selection theory, the heart of which is the concept of inclusive fitness. Kin selection theory is the cornerstone of neo-Darwinian biology, and the mathematical model behind kin selection theory focused on interpreting altruism, or social cooperation.

According to the inclusive fitness concept, natural selection favours cooperation if relatedness, or genetic similarity, is greater than the cost-to-benefit ratio. In Hamilton's formula, the relatedness parameter is expressed by the fraction of genes shared by the altruist and the recipient of altruism.[32] The model, based on the formula, works particularly well in the case of ants because of a specific sex-determining process called haplodiploidy. In this process, fertilised eggs become females and unfertilised eggs become males. The result is that sisters are more related to each other (75% genetic similarity) than mothers and daughters (50% genetic similarity). This means that sterile sisters act as altruists, sacrificing their genes for the good of the colony. In the 1970s and 1980s, kin selection theory and Hamilton's mathematical model dominated the fields of population genetics and sociobiology.

That is, until E.O. Wilson began to discover cracks in the theory. A decade or so after *Sociobiology* was published, Wilson realised that kin selection theory does not work in the case of termites because of a different sex-determining process. Compared to ants, in which haplodiploidy is the consequence of only one reproductive worker – the queen, which gets a lifetime supply of sperm from the nuptial flight stored in an organ called the spermatheca – in termites both the queen and the king are sexually active, leading to diploidy as the sex-determining process (diploidy is also the sex-determining process in humans). According to Wilson and his collaborators, '[t]he association between haplodiploidy and eusociality fell below statistical significance' when they applied Hamilton's model to termites.[33] With time, further evidence accumulated against kin selection theory in the context of eusociality. This prompted Wilson to search for a more suitable mathematical description of eusociality.

Working in collaboration with two Harvard mathematicians, E.O. Wilson developed a new mathematical model of eusociality that does not fully discard kin selection theory but favours group selection as the cornerstone of eusociality. According to the new theory, eusociality develops in five stages: (1) the formation of groups, (2) the emergence of a defensible nest, (3) the emergence of genes that prevent social dispersal, (4) the emergence of traits shaped by the environment and (5) the emergence of multilevel selection that shapes the structure and function of the colony.[34]

If we now go back to the relationship between eusociality and agriculture/engineering, it turns out that group selection theory, but not kin selection theory, can incorporate the development of this relationship. Stages 2, 4 and 5 are particularly important. For example, one can envisage that engineering skills are important, but not essential, for the construction of a defensible nest (stage 2). These skills range from fairly primitive ones in the case of small nests located inside stone cracks or acorn shells, which only require an active defence, to more sophisticated ones that require the construction of the nest from scratch and subsequent defensive strategies.[35] The simultaneous development of construction and agricultural skills takes time and is likely to occur in stage 4. For example, stage 2 incorporates 'massive provisioning', whereby the young queen provides enough paralysed pray for the developing colony when starting to construct the nest. But stage 2 is progressive, meaning that other forms of massive provisioning, or production of food on a large scale, including agriculture, will emerge in subsequent stages of eusociality – for example, in stage 4. Finally, in stage 5, colony selection emerges – a process of sharing environmental resources. A typical leaf-cutter ant colony consumes a quantity of green biomass equivalent to the consumption of an average cow.[36] We can conclude that eusociality, at least in the case of social insects, incorporates the progressive and simultaneous development of engineering and agricultural skills.

However, the reactions of some biologists to this new mathematical model of eusociality and group selection theory were negative. The journal *Nature*, in which Wilson's article was published, received several letters from proponents of kin selection theory, arguing that group selection theory is wrong.[37] The most vocal critic was Richard Dawkins. He reviewed E.O. Wilson's book *The Social Conquest of Earth*, a popular summary of group selection theory, in the most vitriolic terms in *Prospect* on 24th May 2012. Wilson reacted on the BBC programme *Newsnight*. He said that he did not have any dispute with Richard Dawkins 'because [Dawkins is] a journalist, and journalists are people who report what the scientists have found'.[38] E.O. Wilson invited his critics to disprove the mathematical model of eusociality. More than a decade later, this has not happened.

What Makes Us Farmers?

Even though the new group selection theory and the five stages in the development of eusociality are intended to be universal for all eusocial species, E.O. Wilson wanted to avoid the backlash he experienced in 1975 after the publication of *Sociobiology*. Critics accused him of interpreting human nature as the continuum of brutal animal behaviour. To answer the critics, E.O. Wilson published *On Human Nature* in 1978, for which he was awarded the Pulitzer Prize.[39]

Since the new theory of eusociality abandoned kin selection theory that was so dominant in *Sociobiology*, E.O. Wilson thought a new book was warranted to explain group selection theory in the context of human nature. This new book, *The Social Conquest of Earth*, was published in 2012 to demonstrate that we humans are eusocial species, but that stages 4 and 5 'occur only in insects and other vertebrates'.[40] The uniquely human social condition is achieved through human culture. To add philosophical depth to his thinking, E.O. Wilson published another book, *The Meaning of Human Existence*, to argue that '[o]nly wisdom based on self-understanding, not piety, will save us.'[41] Presumably he meant from the pitfalls of our nature.

What is human nature, and what is unique in human nature that makes us a eusocial species, so heavily reliant on agriculture as the basis of modern human civilisations? E.O. Wilson describes human nature in the following way: 'Human nature is the inherited regularities of mental development common to our species. They are the "epigenetic rules" that evolved through the interaction of genetic and cultural evolution that occurred over a long period in deep prehistory.'[42]

The key phrase in the above description is 'epigenetic rules'. According to biology textbooks, epigenetics is the study of phenotypes that do not include changes in the DNA sequence. In other words, epigenetics discards gene mutations as the only causative factor in phenotype changes. This possibility opens the door to the principles of autopoiesis as factors shaping human nature. Organisms are more important than genes. Organisms are governed by autopoietic principles that reduce genes to the biological substrate, malleable by autopoietic principles (epigenetics), but that maintain the importance of genes as a form of biological memory that keeps organisms connected to evolution as the continuous process of life. We can say that human culture is the product of coevolution between autopoietic principles and malleable biological substrates, including biological memory.

We can also detect some parallels between the above formulation of human nature and the concept of biocivilisations, or the cognitive spaces of individual species, translated into biological substrates. Each cognitive space is the consequence of the autopoietic nature of organisms as independent agents. The society of independent agents that belong to the same species makes up the culture of that species. Just as we humans have a unique nature and culture, so do bacteria, elephants, amoebas and all other species (see Chapters 4 to 9).

In Chapter 1, I described bacterial memory, but a more apt description may be bacterial nature. Bacterial nature is a combination of (a) biological information stored in bacterial genomes (genetics) and (b) information the bacterial society collects from the environment and stores in the structure of the society (culture). This genetically stored information is not sufficient for the survival of a bacterial society; it serves only to trigger more complex collective information-processing abilities, or bacterial culture, that will allow bacterial societies to learn about new conditions in the environment, and by doing so propagate bacterial culture.

What are the features of human nature that make our culture unique, including the capacity to develop our version of agriculture? E.O. Wilson argues that our propensity for long-term memory and the capacity for abstract thinking, including the creation of scenarios mimicking real life, combined with forward planning, enabled group selection.[43] Groups with cooperative members capable of using the above behavioural features to their advantage, in competition with other, less gifted groups, were favoured by natural selection. The winning groups developed sophisticated cultural features, including social intelligence based on complex language, elements of morality, religion, artistic flair, fighting abilities, capacities for technical intelligence and natural history intelligence, which were important for the emergence of agriculture.

The development of human culture can be demonstrated using comparative biology. One can compare *Homo sapiens* with related species to reveal their evolutionary trajectories. As a species, we have a lot in common with *Homo neanderthalensis*, or Neanderthals, including some of their genes. The last Neanderthals lived approximately 30,000 years ago. Comparative biology revealed that *Homo sapiens* and Neanderthals had similar brain sizes and similar general intelligence. However, the culture developed by Neanderthals was poor relative to the culture of *Homo sapiens*.[44] The key reason was the inability of Neanderthals to develop more sophisticated types of intelligence, including social intelligence, technical intelligence and natural history intelligence. By contrast, *Homo sapiens* was able to use long-term memory and link it with different domains of intelligence. This led to the development of sophisticated language and the emergence of complex abstract thinking, including metaphors and analogies.

The consequence of cultural differences between *Homo sapiens* and Neanderthals was that Neanderthals remained hunter-gatherers with a primitive and stagnant culture, while *Homo sapiens* discovered agriculture. This, thanks to its economic power, catapulted certain human cultures into powerful civilisations that emerged and fell over the last ten millennia, culminating in modern-day globalised civilisation.

According to an interesting hypothesis, agriculture was a by-product of misplaced social intelligence. Archaeologist Steven Mithen developed the concept of cognitive fluidity.[45] This is a modular human mind that easily connects all domains of intelligence – general, social, technical and natural history intelligence – to create original thoughts based on complex language, the use of metaphors, analogies and inner speech. The origin of cognitive fluidity in *Homo sapiens* can be traced to roughly 70,000 to 50,000 years ago. One feature of cognitive fluidity was the widespread worship of nature by hunter-gatherer societies, as a result of developing natural history intelligence. Our ancestors interpreted plants and animals as relatives. This tradition survives until the present day in some traditional societies.

Given the form of social intelligence that extends from human societies to the entire biosphere as one giant society of related individuals, archaeologist Nicholas Humphrey argued that this, by modern standards, 'fortunate misapplication of social intelligence' created favourable conditions for the emergence of agriculture in hunter-gatherer societies roughly ten millennia ago: 'The care which a gardener gives to his plants (watering, fertilising, hoeing, pruning etc) is attuned to the plants' emerging properties. True, plants will not respond to ordinary social pressures (though men do talk to them), but the way in which they give to and receive from a gardener bears, I suggest, a close structural similarity to a simple social relationship. If [we] can speak of a conversation between a mother and her two-month-old baby, so too might we speak of a conversation between a gardener and his roses or a farmer and his corn.'[46]

Our hunter-gatherer ancestors understood intuitively that human sociality is part of the giant social collective of the biosphere. They saw themselves as the children of forests and prairies, and scenarios emerged justifying the use of brothers and sisters – plants and animals – to keep the kinship alive far into the future. This kinship is the symbiotic nature of agriculture so vividly described by Nicholas Humphrey. Steven Mithen used the concept of cognitive fluidity to support Humphrey's hypothesis. He analysed archaeological evidence at early agricultural sites in the Near East and concluded that, indeed, human sociality in its primordial form – nature as one big family – contributed to the emergence of agriculture.[47]

Going back to E.O. Wilson's elaboration of human nature, we can conclude that the coevolution between genes and culture incorporates agriculture as the essential component of human culture. Without the economic power of agriculture, there would be no civilisation as we know it. Equally, without eusociality and group selection, there would be no agriculture.

A Terribly Botched Job of Engineering

The first message of E.O. Wilson's *The Social Conquest of Earth* is that eusocial insects, such as ants, termites and honeybees, dominated the planet for fifty million years. Their societies were the most complex and ecologically successful in the animal kingdom. Top-down and bottom-up ecological forces constrain the biological potential of leafcutter ants, for example, and keep their biomass in tune with the ecological balance.

His second message is that our eusociality is disastrous for the biosphere. In only a few centuries, our species has become a geophysical force. While the agriculture that emerged at several locations on the planet in the early Holocene was not initially the source of ecological disturbance, it did lead to economic surplus, which in turn fuelled powerful new technologies aimed at harnessing the natural energies of the Sun and fossil fuels. This, of course, led to widespread contamination of the planet with the by-products of these technologies, ranging from concrete and plastics to radioactive particles from nuclear testing and chicken skeletons left over from feeding billions of people.

E.O. Wilson invoked imaginary extraterrestrials to explain our ecological predicament. Scientists from extraterrestrial civilisations far more advanced than our own, and far more experienced in analysing life-like phenomena on the cosmic scale, would be extremely critical of our technologies, including agriculture. They would simply conclude: 'It's a terribly botched job of engineering'. They would even criticise their own responsibility. As a more advanced civilisation that encountered similar problems in their evolutionary past, they should have anticipated our predicament and warned us in time: 'We should have come here earlier and stopped this tragedy from happening.'[48]

You may recall from Chapter 6 that, in the context of biocivilisations, engineering and technology are the same phenomena. While all other species orient themselves to the world in an autopoietic way – natural engineering is regulated by the decentralised Gaian mind – we humans are slaves to our nature, which forces us to interpret the biosphere as territory that must be conquered by our mechanised technologies. Mechanised technologies, in the context of the biosphere, are little else than terribly botched engineering jobs. The failure to distinguish between the organic and the mechanical, the organism and the machine (see Chapter 2), and to acknowledge that engineering machines and modifying nature are fundamentally different things, makes our predicament even worse.

Is there any way out of this terrible predicament? I shall address this question in the final part of the book.

Part III

LOOKING FORWARD

CHAPTER ELEVEN

How to Swim in the River of Life?

Life is a planetary-level phenomenon and Earth's surface has been alive for at least 3,000 million years. To me, the human move to take responsibility for the living Earth is laughable – the rhetoric of the powerless. The planet takes care of us, not we of it. Our self-inflated moral imperative to guide a wayward Earth or heal a sick planet is evidence of our immense capacity for self-delusion. Rather, we need to protect us from ourselves.

LYNN MARGULIS[1]

Our journey through the elusive landscape of biocivilisations is coming to an end. What remains is to make sense of the journey in the age of an increasingly mechanised world view dominated by a single species. The panoply of terms that describe the mechanised outlook includes 'Anthropocene', 'Machinocene', 'technological singularity', 'Industry 4.0' and 'surveillance capitalism'. If there is a single word that effectively describes the mechanised outlook, it must be 'autocracy'. *Homo sapiens* has become a self-styled autocratic ruler of the Earth. The age of man, or the Anthropocene, in the context of Gaia, is akin to a demand or ultimatum issued to all other species and their biocivilisations to voluntarily accept the terms of our rule.

But the brave new world of biocivilisations is a democracy – there are no privileged species. Lynn Margulis immediately spotted this, and she was right. We need to protect ourselves from our own autocracy.

In this chapter, I shall outline five arguments that will help us see ourselves in the true light. I hope that the arguments will not be interpreted as anti-modernist, anti-technological or anti-scientific. As I aimed to present an alternative view of the biological world in the previous ten chapters, I am obliged to present the future outlook based on this alternative view. My motives are the same as the motives of scientists who argued that biology is fundamentally different from physics (see Chapters 1 to 3). If we continue to turn a blind eye to further attempts at mechanising biology, *Homo sapiens* will increasingly resemble 'a vain emperor who just loved clothes. More than ruling his land wisely, he seemed interested only in changing into a new suit every hour.'[2]

A New Science of Life

What would a new science of life look like – one in which biology is freed from the grips of mechanism? How *can* we understand our place within the river of life? With improved understanding and a new course of action, perhaps it is not too late to avoid the violence of Gaia's flow that has been caused by our poor swimming skills. Let's remember two important things: Gaia regulates the flow of life through its decentralised autopoietic mind, and our position within the river of life is insignificant; we are a tiny stream in a huge river.

The first step in generating the curriculum of a new science of life is one of integration and adjustment, and it is best illustrated by comparing two figures: the autonomous circles of Figure 5.1 (page 81) and a new integrated structure illustrated by Figure 11.1. Despite extensive integration and sharing, the circles in Figure 11.1 remain semi-independent domains, but they also take on new dimensions and depth:

The Theatre (T); the watchers and the watched: Discovering our place in the biosphere through watching the biotic theatre with the help of evolutionary retrodiction.

The Workshop (W); naturalised engineering: Learning how to adjust our instrumentalised technologies to the autopoietic nature of the biosphere.

The School (S); naturalised science: Learning how to adjust our increasingly mechanised science to the autopoietic nature of the biosphere.

The Clinic (C); naturalised medicine: Learning how our health depends on the health of other species and the dynamics of Gaia.

The Gallery (G); naturalised art: Discovering the limits of traditional science and liberating it through biotic art.

The Kitchen (K); saint and sinner[3]: Learning how to adjust our biomass to the capacity of planetary resources.

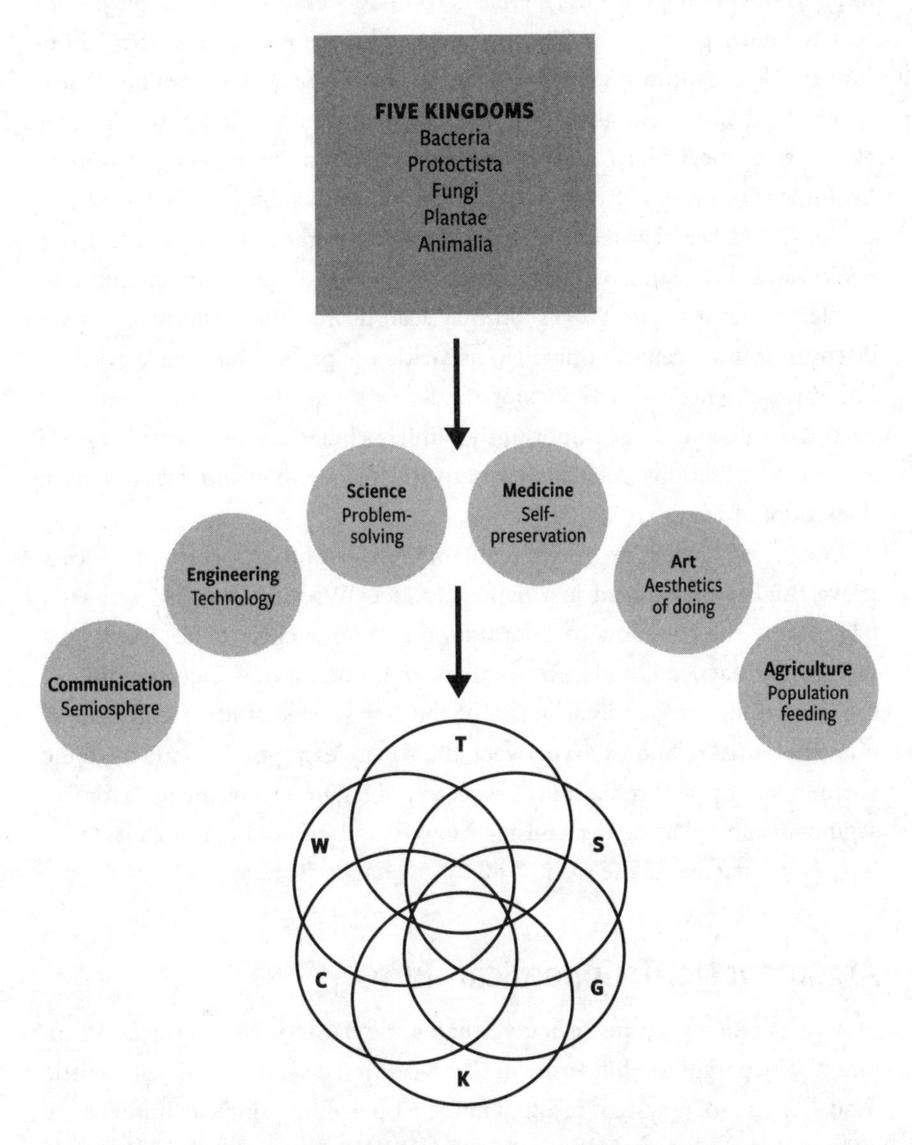

Figure 11.1. Integrating components of biocivilisations represented by six circles into a composite circle dominated by extensive interactions. Abbreviations T, S, G, K, C and W are explained in the text.

Information, ideas and resources now flow freely from one circle to all others, and vice versa. For example, endless feasts that typify the Kitchen circle are historically part of the planetary dynamics or the Theatre circle. Yet we cannot watch the plays in the Theatre without the Kitchen's resources (akin to buying popcorn in cinemas). To understand the various plays constantly taking place in the Theatre, we need to be educated in the School (universities, institutes, secondary schools). In the School, we not only learn about the Theatre but also about the Clinic. Inside the Clinic, we discover elements of the Theatre and also get introduced to the meaning of human health in the context of the environment and the stability of the biosphere. In the Clinic, we also learn that the activities in the Workshop affect our health and the health of many other organisms. Finally, our attempts to tackle problems in the Workshop may lead us into the Gallery, where we discover how art can not only help us tackle the problems in the Workshop but also influence the activities of the School, the Theatre, the Clinic and even the Kitchen. Most importantly, similar circles are regular features in many biocivilisations (see Chapters 5 to 10). We are the youngest student in the school of life.

Integrated circles are reminiscent of Steven Mithen's concept of cognitive fluidity introduced in Chapter 10 (see 'What Makes Us Farmers?', page 190) – the free flow of information from one circle to all others, and vice versa, may lead to discoveries that can influence our understanding of the river of life profoundly. The task of the new science of life is to explore as many potential relationships between the six circles as possible, and by doing so improve our swimming skills. I wish to use Figure 11.1 as the basis for five arguments aimed at demechanising biology and improving our capacity to integrate ourselves in the river of life.

Argument No. 1: Copernican Turn

Biology is ripe for a transformative change similar to the Copernican Revolution. The paradigm shift from an Earth-centred cosmos to the heliocentric model of the solar system resulted in the Copernican principle: humans are not privileged observers of the universe.[4] While the Copernican principle is accepted at the level of the physical universe, it is largely ignored at the level of the biological world. It is difficult for us to accept the fact that we are not superior observers. All organisms are observers of the universe in their way. Most importantly, our observational skills have not passed the test of time (see Figure 3.1, page 48). If we are true believers in the Copernican principle, the Anthropocene should be interpreted as an insignificant branch of the Bacteriocene. It may become a significant branch only if it endures. Even if it does, it will remain a tiny branch growing from a big tree.

Most of us have a deep feeling of cosmic loneliness. Enrico Fermi summed this up by asking a simple question: Where is everybody?[5] All predictions indicate that extraterrestrial civilisations must exist, and yet we do not see them. Why are we alone in the universe? One answer may be that intelligent civilisations destroy themselves prematurely.

But there is an alternative explanation, tantamount to the Copernican turn in biology. It is about the way we perceive the biological world. We are self-centred and too focused on our own intelligence. This makes us blind to the intelligence of our planetmates. Every single organism, from a bacterium to an elephant, is intelligent in its own way. We don't appreciate the intelligence of our planetmates; intelligent extraterrestrials, if they ever visit us, just might.

Indeed, extraterrestrials visiting our planet might conclude that the most intelligent form of life on Earth is microbial. Microbes created the planetary web three billion years ago and their diversification generated other forms of life, which were inferior to microbes by being sensitive to periodic extinctions. These forms of life, including plants and animals, depend on microbial ecological waste. The mode of living of plants and animals is the symbiotic dependency on microbes. Microbes have succeeded in creating the Internet of Living Things, the most recent invention of which is the cosmic adventure by an organism aptly called the thinking ape.[6]

Very few people take microbes seriously, but the thinking ape as a microbial vector, helping the expansion of life from Earth into the cosmos, is a perfect biological scenario. There is no supernatural force involved. Everything is based on autopoietic principles. Microbes have the power of controlling their hosts, including the thinking ape, through the 'microbiota–gut–brain axis'. Microbes create conditions for their expansion by manipulating the hosts.

We simply cannot dominate microbes. This is because there are many more microbial generations, which soak up all sorts of experiences through the Internet of Living Things that we do not appreciate yet. As soon as one microbe develops resistance to an antibiotic, the information is conveyed to the entire planetary microbiota.[7] The biocivilisation of microbes is a biogenic net that has dominated Earth for the last three billion years. We are medium-sized animals born out of this net a few moments ago and will remain entangled in it as long as we live.

Argument No. 2: The Creative Nature

Artists do not need the scientific method to critique the Anthropocene; they rely on artistic instinct. Here is how writer Milan Kundera depicts the predicament of human nature that stems from misunderstanding biology: 'True human goodness, in all its purity and freedom, can come to the fore only when its recipient has no power. Mankind's true moral test, its fundamental test (which lies deeply buried from view), consists of its attitude towards those who are at its mercy: animals. And in this respect mankind has suffered a fundamental debacle, a debacle so fundamental that all others stem from it.'[8]

Kundera does not exaggerate about the 'fundamental debacle' of humanity. Scientists have gathered enough evidence to show that humanity indiscriminately kills other living beings in an act of unprecedented biological pandemonium.[9] We are a biocidal HIPPO that terrorises the planet and yet cultivates the hope of a new deity.

Other artists invented more subtle narratives to challenge our distorted picture of the biological world. Poet Vasko Popa had a talent for seeing 'through the clutter of our civilized liberal confusion'.[10] In the short poem 'Horse', he uses this animal, so dear to us, as an understated symbol of the suffering of the entire animal kingdom at the hands of our superior civilisation.

Horse
Usually
He has eight legs

Between his jaws
Man came to live
From his four corners of earth

Then he bit his lips to blood
He wanted
To chew through that maize stalk
It was all long ago

In his lovely eyes
Sorrow has closed
Into a circle
For the road has no ending
And he must drag behind him
The whole world[11]

Kundera connects with Popa through Nietzsche, a well-known critic of human culture. The link between the trio of artistic and philosophical minds is the horse. Kundera also turns into a psychiatrist. He walks the tightrope between human sanity and insanity.

> *Another image also comes to mind: Nietzsche leaving his hotel in Turin. Upon seeing a coachman beating a horse with a whip, Nietzsche went up to the horse and, before the coachman's very eyes, put his arms around the horse's neck and burst into tears.*
>
> *That took place in 1889, when Nietzsche, too, had removed himself from the world of people. In other words, it was at a time when his mental illness had just erupted. But for that very reason, I feel his gesture has broad implications: Nietzsche was trying to apologise to the horse for Descartes. His lunacy (that is, his final break with mankind) began at the very moment he burst into tears over the horse.*[12]

Kundera's message is devastating: we can acknowledge our biocidal propensity only by going insane. Yet in the world of our collective sanity, we see ourselves as the creatures closest to God, which calls into question our collective sanity. If we believe in human sanity, then our god can only be a monster – a greater monster than HIPPO. Who's responsible for this predicament? Kundera has no doubts. It must be techno-scientific hubris, symbolised by the French philosopher and mathematician René Descartes. For Descartes, animals were mechanical creatures, machines with no souls.

But our hunter-gatherer ancestors saw animals and nature differently. The horse was their close relative, one who could cry, laugh and feel like any other creature with a soul.

If Kundera is correct, the human predicament lies in our mechanical minds, programmed by the dominant techno-scientific outlook. One way to counter the mechanistic outlook is to view nature in its primordial form – a creative system that defies mechanism – and try to adjust our science to the new outlook.

Some scientists have already started the process of refining science to take account of life as a non-mechanical system. For example, Stuart Kauffman argues, contrary to conventional science, that nature is all about creativity that goes beyond natural laws.[13] According to Kauffman, we need to drop the idea of fixed and permanent natural laws and open our minds to the idea of nature as a creative system operating without the constraints of fixed laws. As nature changes, the laws, in the form of various constraints, change too.

Science needs a fundamental change to adjust to a reality without fixed and permanent laws. Kauffman also argues that our hopes for explaining nature fully, through the so-called Theory of Everything, are doomed. If nature is a creative system, conquering it through mechanised science makes no sense, because the creative system cannot predict its next state. If this is so, relying on reason alone – the supreme virtue of enlightenment – is not sufficient for solving the problems we face in our everyday lives. We have to rely not only on reason but also on emotion, intuition, allegory, metaphor and all the linguistic and psychological features that make us humans.

Accepting the truth that *Homo sapiens* is not the supreme species – the Copernican turn in biology – is extremely painful. We have to relinquish the power of presiding over the entire biological world and embrace our role as an insignificant part of that world – a world in which we are early learners who have only just started our schooling. The success of this schooling is measured by evolutionary maturity – an ideal that is far, far away from our reach.

Perhaps even more painful is that we thought we could read the mind of God (the creator of the universe and natural laws), and we also thought we could become God-like ourselves by ruling the rest of God's creations. But this may be a case of monumental misunderstanding. We are poor students. God may not be what we think it is. It may not be the supreme creator of the world and the inventor of laws. According to Stuart Kauffman, nature – life and the planet – is the new sacred. Kauffman asks a question that links him with Kundera's notion of the human debacle: 'How dare we destroy products of this fully natural God?'[14]

Argument No. 3: Mind Is the Essence of Being Alive

If nature is a creative system, it must have a mind. Without the mind, all the products of the creative system, as well as the system itself, would become chaotic to the point that the system would struggle to move from one state to another in a stable manner, and would cease to exist within a relatively short period. However, judging by (1) the lifetime of the system, which is roughly four billion years old (close to one-third of the age of the universe) and (2) the trajectory of the system, or the movement from one state to another without losing the properties of the system, we must conclude that the system is stable by evolutionary standards and that it will remain stable in the future, with or without us.

The process of life is necessarily a cognitive or mind-like process. As we remember from Chapters 1 and 2, the main proponent of the mind as an essential property of life was Gregory Bateson. Bateson's elaboration of the natural mind served as the basis for my concept of hyperthought – the decentralised and distributed mind of the biosphere established through the unity of interacting or communicating parts. In Chapter 5 we saw that communication between members of the biosphere – all the species that make it up – is extensive. These interactions lead to the 'transformation of previous events', or structural changes in the biosphere. Transformations range from naturalised engineering, science and medicine, to naturalised art and non-human agriculture. Brains are not required for communication and transformation. The most dominant groups of species in the biosphere – bacteria and plants – are intelligent without having brains.

Bateson wasn't the only person who viewed nature as a mind-like system. Anaxagoras, a pre-Socratic philosopher, had two original ideas. The first was that 'part of everything is in everything'.[15] Anaxagoras believed that there must be a basic element of every material thing. He called this basic element the 'seed'. The human body, for example, must have seeds for skin, muscle and blood; there are even seeds for dark or light hair. If there is a multitude of seeds, how can we explain the unity of the human body? Or the unity of the cosmos? The explanation represents Anaxagoras' second idea, the idea of the 'cosmic mind' or *Nous* (νοῦς).[16] *Nous* arranges and intelligently unifies all things. Anaxagoras believed that plants also possessed a mind. Charles Darwin and some modern botanists would agree with him. Anaxagoras was the first to use the term 'panspermia' – the idea that life in the form of 'seeds' is scattered throughout the cosmos. Two and a half millennia later, Svante Arrhenius, Francis Crick, Fred Hoyle and Chandra Wickramasinghe introduced the idea of panspermia into science.

After Anaxagoras, only rare philosophers ventured into the elusive landscape of the natural mind. One of them was Charles Sanders Peirce. For him, 'all matter is really mind'.[17] Peirce's idea of triadic semiology introduced the natural mind's proper place within natural communication systems. The first biologist who seriously grappled with the idea of the natural mind was Jacob van Uexküll. In his essay 'A Theory of Meaning', Uexküll intimates that while he was listening to St Matthew Passion in a Hamburg church, he asked himself a question: 'Why should the powerful drama of Nature, which unfolded since the appearance of life on our Earth, not be one single composition in its heights and depths, just like the Passion?'[18]

Given the religious tone of the musical piece, and Uexküll's prior use of the term 'destiny', he possibly appears to endorse creationism. However, he offers a way out of the creationist trap. Later in the text, Uexküll qualifies his understanding of God. For him, it seems, God is not the Bible's version of God. He used the expression 'God-Nature'. One can argue that Uexküll's God-Nature is reminiscent of Spinoza's God. Thus, Uexküll may not be a genuine creationist. Some scientists, including Albert Einstein, endorsed Spinoza's treatment of a deity summarised as *Deus sive natura* (God or nature).

To give Uexküll the benefit of the doubt, one can argue that he was unable to explain scientifically the complexities of nature that translate into its vast mind. Simply put, Uexküll did not know enough, not because he was incapable of knowing, but because biology, in his time, was a young discipline subservient to physics and chemistry. Since Uexküll was not happy with the concepts of natural selection and the survival of the fittest as the organising principles behind the biological world understood as mind, he resorted to God-Nature as the 'cry for help' of a biologist at pains to understand nature's narrative.

If God-Nature was the cry for help launched into the future to rescue a vision of biology from the philosophical wilderness, then the founders of biosemiotics failed at reading Uexküll's subtle symbolism. They remain silent about God-Nature. Instead, rescue came from Stuart Kauffman. The all-powerful creator – God-Nature, the controller and owner of all the matter that exists in the universe – is transformed into the process of 'the ceaseless creativity we can never foretell' that carries us and all other species in its flow.[19]

In point of fact, it was not Kauffman who rescued Uexküll (not to mention Anaxagoras and Peirce) from the constraints of his thinking, but Gregory Bateson. (This is not to say that Kauffman's thinking is not scientific. On the contrary, Kauffman is probably the greatest living proponent of the science of life freed from mechanism.) By combining cybernetics and biology, Bateson liberated the thoughts of his predecessors from theological constraints. Scientists are no longer hesitant to investigate mind-like phenomena previously deemed unscientific: bacterial cognition, plant intelligence, cognition-based evolution, bacterial linguistics, the language of whales, the mathematics of bees and many others.[20] One is even tempted to go beyond the biotic world to search for the origin of the mind; the structure of the human brain is remarkably similar to the structure of the galaxy.[21]

In their book *Into the Cool*, Eric Schneider and Dorion Sagan describe the antecedents of biological selves or physics' own 'organisms': non-living complex systems, Bénard cells, Taylor vortices and Belousov–Zhabotinsky chemical reactions. The natural purposiveness of these organisms to reduce ambient gradients leads to mind-like behaviour that tricked Jesuit priests who attended a chemistry lecture into mistaking them for real organisms.[22] These organisms of physics do not require natural selection to emerge. W. Ford Doolittle and Richard Dawkins, defenders of neo-Darwinism, dismissed Gaia because they could not imagine the biological world without natural selection. But James Lovelock's Daisyworld computer model shows that growth within a temperature gradient is enough to induce mind-like behaviour. The origin of complexity, including the mind, is the consequence of energy flow and the capacities of non-living and living systems to reduce energy and other gradients.

Emerging research suggests that the mind is the vital property of nature, and it is not exclusively a human property. Just as we humans have measurable IQ, so too do bacteria.[23] Eusocial insects, usually considered to be Cartesian automata, seem to have personalities and forms of social behaviour similar to our democracy.[24] Human groups, when placed in a competitive cognitive environment, show group intelligence independent from individual IQs.[25] Plants have rich cognitive worlds in the absence of brains.[26] Most importantly, Gaia as a system composed of intelligent and interacting parts – that may be called the interactome as the manifestation of Gaia's decentralised mind – must have the capacity to make decisions, remember previous states and sense its cosmic environment. Even though Gaia has no brain or central processing unit, biospheric growth and regeneration can preserve complexity and Gaia can repair itself after perturbations in ways and at a scale that deserve to be compared, at least, to human thought and ingenuity.

Argument No. 4: Life Is Symbiosis

Evolution, according to mainstream science, is Darwinian. This is an entirely random process, whereby organisms adapt to their environments through various forms of fitness and natural selection. Darwin himself relied on the concept of 'survival of the fittest' – a phrase coined by Herbert Spencer. Darwin interpreted the concept of fitness to mean reproductive fitness. The most successful organisms leave the most progeny. Neo-Darwinists merged Darwin's ideas with genetics and created the concept of inclusive fitness – the idea that long-term survival depends on genes. Individuals with a gene that promotes the reproduction of all organisms with the same gene survive preferentially. This is also known as kin selection theory.

However, Darwinism is limited to the extent that it cannot be used as a universal theory. It applies only to reproductively isolated organisms such as plants and animals. These organisms propagate through vertical gene transfer – the flow of genes from parents to offspring. Most importantly, plants and animals are evolutionary latecomers. They emerged in the last billion years, but life existed for almost three billion years before them. The forms of life that preceded plants and animals – microbes of different sorts – were not reproductively isolated. HGT, a Lamarckian process par excellence, was widespread during the first two billion years of Gaia's history, suggesting that Darwinism must be supplemented with Lamarckism to fully account for various forms of life. This possibility has been acknowledged by prominent scientists, including the late Carl Woese and Stephen Jay Gould. Lamarckism has also been revived by a small group of contemporary scientists and philosophers who view evolution as the combinatorics of different biological scenarios, rather than a single mechanistic explanation forced on the entire landscape of life.[27]

If there is a single principle that can explain all forms of life, it is not the principle that reduces organisms to genes. Instead, it is an integrating principle – the principle that organisms and their fusions integrate into Gaia. This principle is symbiosis, or living together. Symbiosis encompasses endosymbiosis (microbial fusions), mutualism, commensalism, parasitism, predation and competition.

If we view Gaia as a system that constantly creates itself, the method used by the system may be compared, metaphorically, with the method used by an artist. In previous chapters, I used the metaphor of Gaia the self-sculptor. We can also view Gaia as a self-painter, similar to *Drawing Hands* by M.C. Escher, constantly mixing different colours (akin to the simplest organisms, such as prokaryotes) to produce different shades on the landscape of life (composite organisms, such as eukaryotes) and to create large motifs on its emerging multidimensional landscape (biomes as the ecological integration of all species). For example, the cells that we are made up of are chimeras that contain mitochondria – former aerobic bacteria. Plant cells contain photosynthetic organelles, or chloroplasts – former cyanobacteria. Some animals can do photosynthesis – true plant–animal chimeras.[28] Trees rely on symbiosis with fungi to turn forests into 'smart homes' for thousands of species. Life is more a collaborative game between equal partners – viruses, bacteria, archaea, fungi, plants and animals – and less a competition amongst predators whose behaviour is fully determined by genes.

The idea of nature the artist is not an argument for intelligent design – the creator separate from its creation. On the contrary, nature the artist is the natural mind at work – nature inventing itself through symbiosis. Our analytical minds are addicted to the question 'how?' The result of this endless analytics is the reductive materialism that falsely separates us from the universe and turns us into quasi-independent observers. But if we ask the question 'why?' more often, we may conclude, like Kauffman did, that the existence of a creator separate from its creation is a delusion. Life is a process of ceaseless creativity that carries all organisms, including us, in its flow. And the technique for the ceaseless creativity of life is symbiosis. Even bacteria may be viewed as symbionts of virus-like and other biotic elements.[29]

Argument No. 5: Biology First

Finally, the structure of science is wrong. If life is a creative process that carries us in its flow, our attempts to mechanise the human position within the flow and search for the Theory of Everything from the peak of our mechanistic island are doomed to failure. Our mechanising attempts counter the logic of life. Life is a process of ceaseless creativity that precludes mechanism as the driving force.

Yet mechanism dominates the current structure of science. The most fundamental science is physics. Chemistry and biology are derived from physics. The consequence of the 'physics-first' principle in the structure of science is the dominance of reductive materialism in all sciences, including biology.[30] Reductive materialism works well if we assume that the universe is a mechanical system. However, it cannot explain fully non-mechanical systems, such as living systems.

One of the alleged success stories in biology based on reductive materialism is the discovery of genes. For scientists who specialise in genomics, genes control everything.[31] A total of 20,000 or so genes in the human genome are fundamental cogs that control the machinery of our bodies. The explosion of genetic testing offers the army of health enthusiasts the opportunity to discover genes that may make them sick. Techniques such as CRISPR will one day be able to remove disease-causing genes from our genomes. Sperm banks advise potential customers to consider genetic traits that determine biological features of future children by looking at what's written in donors' genes, from food habits to IQ. The gene-based narrative has entered everyday language. You hear football coaches say that the winning mentality is in their players' DNA. Corporation bosses emphasise that hard work is in the genes of the workforce.

The new comfort zone of health sciences is genetic determinism, which stems from scientific determinism and the physics-first structure of science. We seem to fear the unknown – anything we cannot explain with science.[32] But science that is based on a wrong assumption is not only misleading but worthless. Can biology be explained through the physics-first narrative? A growing number of scientists and philosophers question the validity of gene-centred, deterministic biology. Genes are important for understanding biological processes, but they do not play the central role, because organisms are not machines. Instead, organisms, including the simplest prokaryotes, are independent agents underdetermined by genes and the environment.

That the structure of science dominated by physics may be questionable we can glean from arguments put forward by Karl Popper, who thought that the structure of the universe can be split into three parts, each requiring a different explanatory tool.[33] The universe is in part causal. In other words, it is in part a material, mechanical, abiotic system that can be accounted for by classical mechanics. The universe is also in part probabilistic. Classical mechanics must be supplemented by quantum mechanics, in which there is no clear-cut mathematical certainty. To explain quantum phenomena, we have to rely on a form of calculus dominated by probabilities. Finally, the universe is in part open. Here physics, neither classical nor quantum, can help us understand the openness of the universe. Popper interpreted the openness of the universe as a form of creativity inherent in the process of life.

Given that we live in the biological world – Popper's open universe – and that we are heavily insulated from the mechanical abiotic world – Popper's causal universe – because microbes and plants created a smart biological home for us, we have to question the physics-first principle in the structure of science. In other words, why would the Theory of Everything that explains details of the causal universe be useful for us if we do not understand the biological world?

Scientists are reluctant to change the structure of science for various reasons.[34] But the failure to understand the biological world, even if we can understand the abiotic universe, brings into question the structure of science. Scientific evidence suggests that the main existential threat to humanity is humanity. We have to question our capacity to fully understand our nature as part of the biological world. Robert Rosen and Walter Elsasser exposed the failure of science based on the physics-first narrative to appreciate the special status of biology relative to physics. Rosen indirectly suggested that the most fundamental science is biology rather than physics.

Ceaseless Creativity

To persist, Gaia must constantly create itself. The scientific term that best describes Gaia's fluid persistence is homeorhesis – the steady flow of a system around moving set points. Homeorhesis also means the ceaseless creativity of life, to use the vocabulary of Stuart Kauffman. Many scientists erroneously interpret the Gaian system as a homeostatic system. Lynn Margulis and James Lovelock made the same error in the early days of the Gaia hypothesis. However, Margulis realised that homeostatic systems preclude creativity. Homeostasis is steady state. In other words, homeostatic systems maintain only one state and cannot be creative beyond self-maintenance. If all systems were homeostatic, there would be no evolution. The endless creativity of evolution is the consequence of the ceaseless creativity of life – Gaia is the homeorhetic system par excellence.

The five arguments presented above are centred around the distinction between life as a mechanical phenomenon, fully explicable by physics, and life as the process of ceaseless creativity that requires a new school of thought to liberate biology from the grips of mechanism. The logic of life is not compatible with mechanism because life is a process of ceaseless creativity. Without its enormous creative potential, life would cease to exist. Persistence in the case of life equals creativity.

CHAPTER TWELVE

Equality in Diversity

Science is essentially an anarchic enterprise: theoretical anarchism is more humanitarian and more likely to encourage progress than its law-and-order alternatives.

PAUL FEYERABEND[1]

In this final chapter, I'll reflect on how biocivilisations might be reconciled with mainstream science. More precisely, I'll search for the common ground between two visions of the world that can make them closer to each other. In the human world of today, dominated by industrialised scientific-mechanistic ideology, biology is a non-autonomous discipline dependent on physics. In the world of biocivilisations, all species have authentic civilisations – a possibility that makes biology part of the universe's evolution. A critic of mainstream biology and an expert on chromosomes, Antonio Lima-de-Faria, defined life and the universe in the following way: 'Life is essentially a process by which the universe divides itself into two parts, one of which confronts and inspects the other. An organism is just one of the mirrors that the universe uses to look at itself.'[2]

If organisms are mirrors of the universe, the biosphere is the composite mirror at the heart of which is symbiosis, as the way of projecting the authentic (more than human) reflection of the universe – just as Arcimboldo's *Flora* (see 'Flora or Mona Lisa', page 32) represents the equality of the parts in the diversity of the whole. The domination of a single mirror distorts the picture into a narcissistic fantasy. To deconstruct the fantasy, by giving biology a more authentic voice, I'll rely on philosophy. This deconstruction may establish a common ground between the two visions of the world, as a seed for future reconciliation.

The philosopher who can help us in the process of deconstruction is Paul Feyerabend. He investigated the history of ideas in physics and described science as a form of epistemological anarchy. Many saw him as a controversial figure, even an enemy of science. However, a detailed look at Feyerabend's philosophy reveals someone who criticised the philosophy of science, not science itself.[3] Feyerabend argued that there is no method in science. Scientists are opportunists in a pragmatic sense. If a dream based on mythology can reveal a link to reality that leads to scientific discovery, that's good for science but bad for the method.[4]

If we expose biology to Feyerabend's philosophical scrutiny, one anomaly becomes apparent. Biology is dominated by a single intellectual framework: neo-Darwinism. This makes it a dogmatic science. A non-dogmatic science is characterised by the diversity of ideas –'anything goes', as Feyerabend puts it.[5] But first, for reasons that will become clear, I'll start with some personal reflections on how this book developed as a project, before I turn to Feyerabend's philosophy.

Personal Perspective

I wanted to write this book for a long time. My main motivation was to integrate neo-Darwinism with other ideas to make biology a more diverse discipline, without losing its scientific credibility. Thirty years ago, however, the realist in me shouted: 'Who are you to preach to your colleagues that they should rethink the way they study nature? The scientific community does not like dissenters and will make a mockery of your story.' That realist was also familiar with examples from the history of science about correct ideas being dismissed by the intransigence of the thought collective.

Thirty years on, the so-called realist in me has been forced to change his tune. The thought collective is no longer a monolith entity. It turned out I was not alone, all these years, in doubting the wisdom of the mainstream science of life. Dissenters were everywhere, and they were quiet for the same reason I was. However, something changed a decade or so ago. It became clear to me that individuals cannot challenge the powerful collective (except for some, like Lynn Margulis), but if a few individuals join forces, the collective cannot dismiss them so easily.

A project called extended evolutionary synthesis (EES) was born around 2010.[6] EES took up ideas first introduced in the 1950s that challenged modern synthesis, the view of evolution dominated by neo-Darwinism. EES, a grouping of dozens of scientists supported by eight universities, and an equal number of non-affiliated scientists, does not aim to replace modern synthesis. Instead, EES aims, modestly, to be 'deployed alongside it to stimulate research in evolutionary biology'.

However, a group of scientists closer to my area of expertise, who refer to themselves as The Third Way, challenges neo-Darwinism in no uncertain terms.[7] The founders of The Third Way, launched in 2014, are James Shapiro, a brilliant microbiologist and geneticist (see Chapter 1), Denis Noble, an equally brilliant Oxford physiologist, and Raju Pookottil, an engineer and entrepreneur. The force of their message – that neo-Darwinism is becoming an obstacle to understanding the science of life – was so powerful that The Third Way has seventy-eight members today. It is called The Third Way because its members do not want to be associated with another powerful ideology: intelligent design. All members of The Third Way are respectable and productive scholars, people who have left their creative mark on the science of life.

With The Third Way as an alternative to the mainstream science of life, I realised I wasn't alone. I may not agree with all their views, nor they with mine, but we share a vision of what the biology of the future should be. This gives me confidence that the story of biocivilisations will find the right home in the emerging landscape of the science of life that informs other sciences.

Against Collectivism

The term 'The Third Way' is nicely symbolic. One part of symbolism is a declaration that the two dominant ways of interpreting biological diversity, neo-Darwinism and creationism, should be questioned. Most scientists, including myself, do not agree with creationism, so this is a disagreement by default. However, questioning neo-Darwinism demands the need for a different theoretical framework in biology. The framework requires 'a deeper and more complete exploration of all aspects of the evolutionary process'.[8] The evolutionary process encompasses a multitude of levels, from biomolecules via simple cells (bacteria and archaea) and more complex cells (cells with a nucleus, or eukaryotic cells) to societies of cells merging into multicellular organisms, ecological collectives and Gaia as the terrestrial biosphere representing the planetary living system. The multitude of levels cannot be reduced to neo-Darwinian genetic determinism – random mutations in DNA – as the exclusive source of biological novelty.

Founders of The Third Way James Shapiro and Denis Noble proposed alternative explanatory frameworks. Noble, a physiologist, used the metaphor of the music of life to advance the thesis that life processes are multitudes of synchronised actions at different levels of biological organisation (cells to ecosystems), akin to complex orchestral music encompassing notes, instruments, players, composers and listeners, all tuned together in a harmony representing the melody of life.[9] James Shapiro was less poetic but no less original. He argued that all biological agents, from bacteria to mammals, actively rearrange their genomes in a process he called the 'read-write genome'.[10] Genomes are not passive repositories of DNA sequences but active 'melting pots' of DNA rearrangements as sources of biological novelty via novel RNA sequences and proteins. If, for Noble, life is music, for Shapiro, life is literature: organisms 'write' the story of life using genomes as 'word processors'.

There is another kind of symbolism in The Third Way: individuality as opposed to collectivism. Each member of The Third Way has a story to tell. The Third Way is also My Way, as in the popular song. Frank Sinatra used to announce the song by saying: 'We will now do the national anthem, but you needn't rise.' The beauty of My Way is that everyone can claim it. One does not need to be a singer of Sinatra's calibre to impress.

A recent analysis by a sociologist of science revealed that there may have been as many as sixty-four forms of unintended and beneficial discoveries, known as serendipity.[11] The diversity of forms behind scientific discoveries reflects the idiosyncrasies of the cognitive process, personal preferences, unintended errors and many other sociological phenomena. British immunologist Sir Peter Medawar claimed that the way research is reported in scientific papers is inappropriate.[12] We ignore sociological experiences in the research process at the expense of the so-called objectivity of the research method, but every discovery is more in tune with the idiosyncrasies of My Way than it is the result of the rule-based process of forcing nature to reveal its inner workings to our rational minds.

Anything Goes

After all, Paul Feyerabend was right. His phrase 'anything goes' is a synonym for the diversity of processes behind scientific discoveries. Feyerabend is probably the only philosopher of science who understood the position of biology in the hierarchy of the sciences. The process of scientific discovery is truly biological. We act as 'mirrors through which the universe looks at itself' – cognitive mirrors that can scan the universe and its history, including our own. Each one of us is a unique mirror. Newton's mind mirrored the universe differently from Einstein's. But there are no perfect mirrors. Feyerabend often said that the best minds of science also entertained wrong ideas. This challenges the almost unlimited confidence of some scientists in AI as a way of solving future problems. If we can't have total confidence in the best human minds, can we have confidence in the human mind's artefact?

We must not forget that there are cognitive mirrors much older than us. Bacteria and archaea have been acting as the universe's mirrors for the last four billion years. Plants, animals, protists and fungi are authentic mirrors too. Finally, the mirrors above bacteria and archaea are not made of pure materials. The most complex mirrors are blends of materials.

The common ground between the two visions of the world is Feyerabend's philosophy. The scientific-mechanistic outlook – in the form of the sociology of science that identified sixty-four forms of unintended and beneficial scientific discoveries – is converging on Feyerabend's position: that the process of scientific discovery is a form of epistemological anarchy. As we humans mirror the universe in a multitude of forms and styles, so do millions of other organisms. What unifies us all is Gaia's hyperthought – a giant mirror made of trillions of pieces, all acting on behalf of the becoming of the universe.[13]

John Horgan, a science writer who interviewed Feyerabend and other philosophers of science, including Karl Popper and Thomas Kuhn, sensed Feyerabend's uniqueness: 'Feyerabend attacked science not because he actually believed it was no more valid than astrology or religion. Quite the contrary. He attacked science because he recognized – and was horrified by – science's vast superiority to other modes of knowledge. His objections to science were moral and political rather than epistemological. He feared that science, precisely because of its enormous power, could become a totalitarian force that crushes all its rivals.'[14]

If we accept the above common ground between the two visions of the world, the mainstream science of life must question its reliance on neo-Darwinism as the main intellectual framework guiding biology through the latest industrial revolution (aka Industry 4.0, according to World Economic Forum executive chairman Klaus Schwab).[15] An authentic industrial revolution must be open to other explanatory frameworks. This also means that we should take seriously our fallibility as cognitive mirrors. As a human collective, we do not have the right to dismiss art, philosophy and other forms of human mirroring of the universe, no matter whether they are epistemologically inferior to science or not.

If finding common ground can stimulate a form of diplomacy leading to reconciliation, the main diplomatic skill will be humility. One way of appreciating humility is to force ourselves to look at the world through the eyes of others. The universe, with life in it, is a true cognitive multiverse. The beauty of human civilisation is its true cosmopolitanism. We are biological conglomerates – associations of viruses, naked genes, bacteria, archaea and complex cells formed by their mergers – that uniquely mirror the universe. But uniqueness does not mean superiority. Quite the contrary. Strangely, uniqueness means equality encapsulated in the sense of wonder, which justifies a limited form of teleology – our desire to know what the future holds. The same 'desire' exists in all other living beings, as well as non-living complex systems, by virtue of the thermodynamic purposefulness of energy distribution.

EPILOGUE

A Report to an Academy

In the short story 'A Report to an Academy' by Franz Kafka, an ape called Red Peter describes a transformative experience: his mental journey from being an ape, free to do whatever he wanted in the jungle (a symbol of the free world), to being an individual captured by humans, one who has suddenly lost his precious freedom and realised that the only way to regain it is to *become* human. Red Peter tells an esteemed audience of academics that he enjoyed the transformation and now appreciates what it means to be a human. He is also honest enough to acknowledge that the experience cost him his identity. He no longer feels like an ape.

Our journey was different. It was not forced. I hope that it was at least thought-provoking. The experience of travelling through the mysterious but open world of biocivilisations was directed at discovering errors not visible in our world. We have every reason to hope that this new experience will make us more human, not less. The experience carries a subtle message: we are not gods and will never be. God is something entirely different from our vision of the absolute. The sooner we realise this, the greater our chances of surviving the next few centuries. Otherwise, Gaia, who is not a deity but the self-referential river, will eliminate the human stream from it.

Several people made this journey possible. Lynn Margulis and Gregory Bateson were inspirational guides on various parts of the journey. Their intellectual legacies can be summed up in the words 'symbiosis' and 'mind'. Robert Rosen is the courageous challenger of the structure of science. Humberto Maturana's and Francisco Varela's concept of autopoiesis is still unappreciated by mainstream biology but is essential to the new science of life. Admirers of bacterial civilisation James Shapiro, Eshel Ben-Jacob, Sorin Sonea and Lynn Margulis are truly far-sighted thinkers whose visions are essential for the art of projecting the future. Finally, Jakob von Uexküll and Stuart Kauffman complemented each other in the vision of nature as 'ceaseless creativity'.

If there is a single word that could unify all the ideas explored in this book and capture the essence of the journey, that word should be able to make the essence of life as we know it distinct from the mechanical outlook of modern civilisation. I hope that I have found a suitable word in the scientific legacy of E.O. Wilson, who acted as our guide in the biocivilisation of ants. That word is 'biophilia'.[1] We are born with the instinct to love anything that lives, and ultimately to love Gaia as the unifier of all living forms. Yet our modern civilisation has instilled in us a love for the antithetical world, or mechanophilia. The secret to our survival is in establishing the right balance between the two types of love. A visionary philosopher, Charles Sanders Peirce, in the spirit of Empedocles, proclaimed 'passionate love and hate [are] the two coordinate powers of the universe'.[2] Let's hope that biophilia will help us heal the wound brought about by mechanophilia and clear up the ecological mess we have made.

Let me finish by paying tribute to Franz Kafka and his art of storytelling. A good story has a life of its own. That is, a story can serve purposes that the author did not intend. For me, Kafka's story about Red Peter and his transformative experience is about the direction of travel in the river of life. Travelling towards the human current and away from those of other species – a journey engineered by humans since the Industrial and Scientific Revolutions – is almost certainly a one-way ticket to not only physical but also cognitive extinction. Loss of identity is a kind of mental extinction. All the species we use in our increasingly mechanised culture will ultimately become Red Peters – autopoietic agents forced to renounce their identities. On the other hand, the journey from the human current towards any other current in the river of life is a truly transformative journey. It has the saving power for humans, but also a message to Gaia and her decentralised mind that the new student in the school of life is moving in the right direction.

Acknowledgements

I am immensely grateful to Chelsea Green Publishing for being willing to publish this book. Jon Rae and Margo Baldwin encouraged me to complete the manuscript after reading the initial chapters. Senior editor Brianne Goodspeed is acknowledged for her enthusiasm and exquisite editorial skills. I am indebted to James A. Shapiro, Dorion Sagan, Nathalie Gontier and Josh Mitteldorf for their insightful comments, which identified various flaws in the first version of the manuscript and helped me rectify them. Finally, I would like to say a big thank you to my wife, Miryana, and my daughter, Maria, for their genuine support, without which this book would not have been possible.

Notes

Introduction: The Mystery of Life

1. One of the main postulates in biology is that all modern organisms have a universal ancestor. Charles Darwin argued that a genealogical tree can be constructed to connect all living organisms, past and present. In 1960, Carl Woese identified a 'universal molecular chronometer', small ribosomal RNA molecules that make up ribosomes, the structures in cells on which protein synthesis takes place. Using this genetic chronometer, Woese discovered a new domain of life, archaea, and constructed a new tree of life consisting of three domains: Bacteria, Archaea and Eukarya. The domain Eukarya contains all plants and animals. What I call 'branches' are domains in scientific vocabulary.
2. Biological information is stored in DNA owing to the genetic code. Each protein is synthesised on the basis of DNA code in which three letters in the DNA 'alphabet' (adenine, guanine, thymine and cytosine) are translated into one amino acid, a building block of proteins. However, this simple genetic code works only in bacteria and archaea. In eukaryotes, epigenetics dominates. Epigenetics means that many phenotypic features can be inherited without changes in the DNA sequence. This is because eukaryotes have chromosomes, structures that suppress or activate DNA sequences selectively.
3. Freeman Dyson, *Origins of Life* (Cambridge, UK: Cambridge University Press, 1999), Kindle location 1117 (emphasis my own).
4. Anne Buchanan and Kenneth Weiss, 'Things genes can't do,' *Aeon* (9 April 2013), https://aeon.co/essays/dna-is-the-ruling-metaphor-of-our-age. The quote is from the subtitle of the essay: 'Simplistic ideas of how genes "cause" traits are no longer viable: life is an orderly collection of uncertainties.'
5. Karl Popper argued in *The Open Universe: An Argument for Indeterminism* (London: Routledge, 1988) that the universe is an open and indeterministic system that will forever remain beyond scientific efforts. See more about this in Chapter 11.
6. Lynn Margulis often said that bacteria are indestructible because they can adapt to any terrestrial habitat. The fact that bacteria have existed since the dawn of life means that they have defied all potential extinction threats.
7. A study published in 2020 revealed the method by which bacteria construct their metropolises. Amauri J. Paula, Geeslu Hwang and Hyun Koo, 'Dynamics of bacterial population growth in biofilms resemble spatial and structural aspects of urbanization,' *Nature Communications* 11 (2020), https://doi.org/10.1038/s41467-020-15165-4.
8. Franz Kafka, *The Complete Short Stories* (London: Vintage Classics, 2018), 269–279.

Chapter 1. How to Build a Biocivilisation

1. This quote is from a personal communication from Gregory Bateson to Fritjof Capra. See Fritjof Capra and Pier Luigi Luisi, *The Systems View of Life: A Unifying Vision* (New York: Cambridge University Press, 2014), 253.

2. The dialogue between Carl Sagan and Ernst Mayr was published in a little-known newsletter: *Bioastronomy News*, 1995 (Vol. 7, Nos. 3 & 4).
3. Nick Bostrom and Milan Ćirković, *Global Catastrophic Risks* (Oxford: Oxford University Press, 2008).
4. James A. Shapiro, 'Bacteria are small but not stupid: Cognition, natural genetic engineering, and socio-bacteriology,' *Studies in History and Philosophy of Biological and Biomedical Sciences* 38 (2007): 807–819.
5. Eshel Ben-Jacob, 'Bacterial wisdom, Gödel's theorem, and creative genomic webs,' *Physica A* 248 (1998): 57–76.
6. The term 'bacteriosphere', or the World Wide Web of genetic information, was used by Sorin Sonea in his books and papers. Here is an example: Sorin Sonea, 'A bacterial way of life,' *Nature* 331 (1998): 216.
7. Eshel Ben-Jacob et al., 'Bacterial linguistic communication and social intelligence,' *Trends in Microbiology* 12 (2004): 366–372.
8. Richard Feynman, a Nobel Prize–winning physicist, had written these words on a blackboard at the time of his death.
9. Gregory Bateson, *Steps to an Ecology of Mind* (New York: Ballantine Books, 1972). The quote is from the essay 'Pathologies of Epistemology', page 483.
10. 'Cells that fire together, wire together' is a summary of the Hebbian theory in neuroscience developed by Donald Hebb, *The Organization of Behavior* (New York: Wiley & Sons, 1949).
11. The process behind the construction of environments by organisms is also known as niche construction. The science behind niche construction is elaborated in extended evolutionary synthesis, an evolutionary theory critical of neo-Darwinism. See, for example, Kevin N. Laland et al., 'Does evolutionary theory need a rethink?,' *Nature* 514 (2014): 161–164.
12. Multicell organisms are ecological communities, or holobionts. See, for example, Lynn Margulis, *Symbiosis as a Source of Evolutionary Innovation* (Cambridge, MA: MIT Press, 1991). In brief, all organisms above prokaryotic microbes are communities that form discrete ecological units.
13. Niall Ferguson, *Civilization: The West and the Rest* (London: Penguin Books, 2011), 3.
14. Bateson, *Steps to an Ecology of Mind*, 482.
15. Lynn Margulis, *Symbiotic Planet: A New Look at Evolution* (London: Basic Books, 1999).
16. William F. Ruddiman et al., 'Defining the epoch we live in,' *Science* 348 (2015): 38–39.
17. Jack A. Gilbert and Josh D. Neufeld, 'Life in a world without microbes,' *PLoS Biology* 12 (2014), https://doi.org/10.1371/journal.pbio.1002020.
18. This is the title of a bestselling book: Yuval Noah Harari, *Homo Deus: A Brief History of Tomorrow* (New York: HarperCollins, 2017). It is based on the wrong assumption that *Homo sapiens* is the only technological species.
19. A good summary of the optimism behind human technological development is given by Ray Kurzweil in *The Age of Intelligent Machines* (Cambridge, MA: MIT Press, 1990). The main prediction, in later books, internet documents and lectures, is that intelligent machines will become 'alive' (undergo an intelligence explosion) in 2045, a concept known as the technological singularity. Machine intelligence, if harnessed by humans, will enable various transhumanist transitions, including the eradication of all diseases and human immortality.
20. Bertrand Russell, *A History of Western Philosophy* (London and New York: Routledge, 2004), 345.
21. Richard Dawkins, *The Selfish Gene* (Oxford: Oxford University Press, 1976), 19.
22. Ohad Lewin-Epstein, Ranit Aharonov and Lilach Hadany, 'Microbes can help explain the evolution of host altruism,' *Nature Communications* 8 (2017), https://doi.org/10.1038/ncomms14040.

23. Kalevi Kull, 'Umberto Eco on the biosemiotics of Giorgio Prodi,' *Sign Systems Studies* 46 (2018): 352–364.
24. Details of bacterial communication, known as quorum sensing, can be found in these papers: James A. Shapiro, 'Thinking about bacterial populations as multicellular organisms,' *Annual Review of Microbiology* 52 (1998): 81–104; James A. Shapiro, 'Bacteria as multicellular organisms,' *Scientific American* June (1988): 82–89.
25. Zohar Erez et al., 'Communication between viruses guides lysis-lysogeny decisions,' *Nature* 541 (2017): 488–493.
26. Interpreting bacterial communication without a coordinating activity of mind, as an emergent phenomenon in a social group, is reflected in the term 'small talk', used to describe bacterial language in the following paper: Bonnie L. Bassler, 'Small talk: Cell-to-cell communication in bacteria,' *Cell* 109 (2002): 421–424.
27. Ben-Jacob, 'Bacterial wisdom, Gödel's theorem,' *Physica A*.
28. Gregory Bateson, *Mind and Nature: A Necessary Unity* (New York: E. P. Dutton, 1979), 92.
29. Predrag Slijepčević, 'Micobes have their own version of the internet,' *The Conversation* (4 August 2017), https://theconversation.com/microbes-have-their-own-version-of-the-internet-75642.
30. Charles Darwin, *Descent of Man, and Selection in Relation to Sex* (New York: D. Appleton and Company, 1889), 126.
31. This is the consequence of vertical gene transfer or the evolutionary tree of life.
32. Laura A. Hug et al., 'A new view of the tree of life,' *Nature Microbiology* 1 (2016), https://doi.org/10.1038/nmicrobiol.2016.48.
33. Karin Moelling and Felix Broecker, 'Viruses and evolution – viruses first? A personal perspective,' *Frontiers in Microbiology* 10 (2019), https://doi.org/10.3389/fmicb.2019.00523.
34. Kurzweil, *The Age of Intelligent Machines*.
35. Huw Price, 'Now it's time to prepare for the Machinocene,' https://aeon.co/ideas/now-it-s-time-to-prepare-for-the-machinocene; Nick Bostrom, *Superintelligence: Paths, Dangers, Strategies* (Oxford: Oxford University Press, 2014).
36. Gerardo Ceballos, Paul R. Ehrlich and Rodolfo Dirzo, 'Biological annihilation via the ongoing sixth mass extinction signaled by vertebrate population losses and declines,' *Proceedings of the National Academy of Sciences, USA* 114 (2017): E6089–E6096 , https://doi.org/10.1073/pnas.170494911; Justin McBrien, 'Accumulating extinction planetary catastrophism in the Necrocene,' in *Anthropocene or Capitalocene? Nature, History, and the Crisis of Capitalism*, ed. Jason W. Moore (Oakland: PM Press, 2017), 116–137.
37. Nick Bostrom, *Anthropic Bias: Observation Selection Effects in Science and Philosophy* (New York: Routledge, 2002).
38. From Chief Seattle's speech around 1854. The original version was modified several times. The cited words are probably part of the modification carried out by scriptwriter Ted Perry.
39. Margulis, *Symbiotic Planet*, 161.
40. 'Pathologies of Epistemology' is the title of an essay in Bateson's *Steps to an Ecology of Mind*, 478–487.
41. SET was originally elaborated in the 1967 paper by Lynn Margulis, then Lynn Sagan, 'On the origin of mitosing cells,' *Journal of Theoretical Biology* 14 (1967): 255–274. Later papers by Margulis enriched SET with new details. Several papers have been published by other authors to interpret SET.
42. See this video: https://www.youtube.com/watch?v=TxsVroiUHik.
43. Humberto R. Maturana and Francisco J. Varela, *Autopoiesis and Cognition: The Realization of the Living* (London: D. Reidel Publishing Company, 1980).
44. The concept of the cognitive multiverse is explained in Predrag Slijepčević, 'Natural intelligence and anthropic reasoning,' *Biosemiotics* 13 (2020): 285–307.

45. The concept of cognitive space is explained in Slijepčević, 'Natural intelligence and anthropic reasoning,' *Biosemiotics*. In brief, a cognitive or semiotic space is a species-specific communicative and epistemic platform. Through exchanging semiotic symbols, individual members of a species communicate and learn.
46. Sagan, 'On the origin of mitosing cells,' *Journal of Theoretical Biology*.
47. Predrag Slijepčević, 'Viruses are part of evolution: An Interview with Professor Chandra Wickramasinghe,' *Epistemology of Nature*, http://www.biocivilizations.com/viruses-are-part-of-evolution.
48. Stuart A. Kauffman, 'Prolegomenon to patterns in evolution,' *BioSystems* 123 (2014): 3–8.

Chapter 2. Against Mechanism

1. Walter M. Elsasser, *Reflections on a Theory of Organisms* (Baltimore, MD: Johns Hopkins University Press, 1998), 3. The full sentence from which this quote is taken reads: 'Since the theoretical parts of all past natural science have been Cartesian in this sense, we may conclude that biology is fundamentally and qualitatively different from physical sciences.'
2. The term 'reductive materialism' was used in Kauffman, 'Prolegomenon to patterns in evolution,' *BioSystems*. Here is a relevant quote from the paper: 'Briefly, we remain children of Newton and classical physics, despite the vast transformation Darwin wrought. Newton then Laplace laid the foundations of modern "Reductive Materialism", which remains our model of science itself, including "Dreams of a Final Theory"' (page 3).
3. Steven Rose, *Lifelines: Life Beyond the Gene* (London: Vintage, 2005), 46.
4. Ceballos et al., 'Biological annihilation via the ongoing sixth mass extinction,' *Proceedings of the National Academy of Sciences, USA*; Justin McBrien, 'Accumulating extinction planetary catastrophism in the Necrocene,' in *Anthropocene or Capitalocene?*
5. Margulis, *Symbiotic Planet*, 82.
6. The percentage of the viral DNA in the human genome has been estimated in the following paper: Moelling and Broecker, 'Viruses and evolution – viruses first?' *Frontiers in Microbiology*. The authors argued that at least 8% of the human DNA has a retroviral origin, possibly more.
7. Hakon Jonsson et al., 'Differences between germline genomes of monozygotic twins,' *Nature Genetics* 53 (2021): 27–34.
8. Camila U. Rang, Annie Y. Peng and Lin Chao, 'Temporal dynamics of bacterial aging and rejuvenation,' *Current Biology* 21 (2011): 1813–1816.
9. Elsasser, *Reflections on a Theory of Organisms*, 24–35.
10. Nicolas Rashevsky was the founder of the school of relational biology. A brief introduction to the school can be found at: https://ahlouie.com/relational-biology.
11. Robert Rosen, *Life Itself: A Comprehensive Inquiry into the Nature, Origin, and Fabrication of Life* (New York: Columbia University Press, 1991), 17, 23.
12. The source of Kauffman's quote is Kauffman, 'Prolegomenon to patterns in evolution,' *BioSystems*, 3. The other quote is attributed to Ernest Rutherford. See Rosen, *Life Itself*, 2.
13. The source of this quote is a lecture Richard Feynman delivered at a conference at Cornell University in 1964.
14. Rosen, *Life Itself*, 13.
15. Daniel J. Nicholson and John A. Dupré, eds., *Everything Flows: Towards a Processual Philosophy of Biology* (Oxford: Oxford University Press, 2018).
16. Alfred North Whitehead used the phrase 'dogmatic common sense' in his criticism of science. Here is a relevant quote: 'Nothing is more curious than the self-satisfied dogmatism with which mankind at each period of its history cherishes the delusion of the finality of its

existing modes of knowledge. Sceptics and believers are all alike. At this moment scientists and sceptics are the leading dogmatists. Advance in detail is admitted: fundamental novelty is barred. This dogmatic common sense is the death of philosophic adventure. The Universe is vast.' From Alfred North Whitehead, *Essays in Science and Philosophy* (New York: Greenwood Press, 1968), 121.

17. The concept of constraints in the context of processes is explained in the following paper: Maël Montévil and Matteo Mossio, 'Biological organisation as closure of constraints,' *Journal of Theoretical Biology* 372 (2015): 179–191. A popular interpretation of this important paper is presented in Stuart A. Kauffman, *A World Beyond Physics: The Emergence & Evolution of Life* (New York: Oxford University Press, 2019), 17–33.
18. Lynn Margulis, 'Kingdom Animalia: The zoological malaise from a microbial perspective,' *American Zoologist* 30 (1990): 861–875.
19. Simon Musall et al., 'Single-trial neural dynamics are dominated by richly varied movements,' *Nature Neuroscience* 22 (2019): 1677–1686.
20. Niels Bohr said this in Copenhagen in 1952. The primary source is Werner Heisenberg, *Physics and Beyond* (New York: Harper & Row, 1971), 206.
21. Predrag Slijepčević, 'Flora or Mona Lisa?,' *Philosophy Now* 133 (2019): 18–19. Reprinted with permission.
22. The estimates of species numbers vary. The one referring to the number of species between one and six billion comes from Brendan B. Larsen et al., 'Inordinate fondness multiplied and redistributed: The number of species on Earth and the new pie of life,' *The Quarterly Review of Biology* 92 (2017): 229–265. However, this estimate includes microbes such as bacteria and archaea. This is contrary to the opinions of many biologists, including Lynn Margulis, who thought that bacteria do not have species because of a prevalent HGT. Only organisms that acquire genes through vertical gene transfer (i.e. plants and animals) can be classified as species.
23. The school of thought that dominates biology today is known as modern synthesis or neo-Darwinism. It combines Darwinian evolution by natural selection and genetics. The criticisms of modern synthesis are outlined in the following article: Laland et al., 'Does evolutionary theory need a rethink?,' *Nature*.
24. The theory of organisms is the concept that was first elaborated by Walter Elsasser (see Elsasser, *Reflections on a Theory of Organisms*). A number of biologists supported the idea in different forms, including Gregory Bateson, Lynn Margulis, Humberto Maturana, Francisco Varela, Robert Rosen, Fritjof Capra and others. One of the key concepts of the theory is that organisms are natural agents (see also Nicholson and Dupré, *Everything Flows*).
25. Nicholson and Dupré, *Everything Flows*.
26. *Purpose and Desire* is a book by J. Scott Turner published in 2018 by HarperCollins. Turner argued that the key feature of all organisms is purpose and desire, and that the concept of homeostasis elaborated by Claude Bernard in the 19th century is behind purpose and desire.
27. Maturana and Varela, *Autopoiesis and Cognition*.
28. Robert Rosen, *Anticipatory Systems: Philosophical, Mathematical, and Methodological Foundations* (Oxford: Pergamon Press, 1985).
29. Rosen, *Life Itself*, 17.
30. J. Scott Turner, *Purpose and Desire: What Makes Something "Alive" and Why Modern Darwinism Has Failed to Explain It* (New York: HarperCollins, 2018), 14.
31. Montévil and Mossio, 'Biological organisation as closure of constraints,' *Journal of Theoretical Biology*.

32. Dorion Sagan and Lynn Margulis, '"Wind at life's back" – Toward a naturalistic Whiteheadian teleology: Symbiogenesis and the second law,' in *Beyond Mechanism: Putting Life Back into Biology*, ed. Brian G. Henning and Adam C. Scarfe (Lanham, MD: Lexington Books, 2013), 205–233.
33. Turner, *Purpose and Desire*, 291–292.
34. Lynn Margulis' theory, SET, was elaborated over a period of several decades, starting with her seminal paper in 1967. The review of SET can be found in Predrag Slijepčević, 'Serial endosymbiosis theory: From biology to astronomy and back to the origin of life,' *BioSystems* 202 (2021), https://doi.org/10.1016/j.biosystems.2021.104353.
35. Margulis, 'Kingdom Animalia,' *American Zoologist*, 863.
36. James Lovelock described the negative reception of the Gaia idea in his book *The Ages of Gaia: A Biography of Our Living Earth* (Oxford: Oxford University Press, 1995).
37. Margulis, *Symbiotic Planet*.
38. Conrad H. Waddington, *The Strategy of the Genes: A Discussion of Some Aspects of Theoretical Biology* (London: George Allen & Unwin Ltd., 1957).
39. Margulis, 'Kingdom Animalia,' *American Zoologist*, 866.
40. Margulis, *Symbiotic Planet*, 148–149.
41. Margulis, 'Kingdom Animalia,' *American Zoologist*, 866.
42. Kauffman, 'Prolegomenon to patterns in evolution,' *BioSystems*.
43. Bateson, *Mind and Nature*.
44. The most suitable popular books summarising Lovelock's and Margulis' work are Margulis, *Symbiotic Planet* and Lovelock, *The Ages of Gaia*.
45. In a recent paper F. Vazza and A. Faletti, 'The quantitative comparison between the neuronal network and the cosmic web,' *Frontiers in Physics* 8 (2020), https://doi.org/10.3389/fphy.2020.525731, Vazza and Faletti compared the structural organisation of the neuronal network in the brain and the cosmic web. The results showed remarkable similarities in the network configurations, despite the difference in spatial scale. The authors concluded that remarkable similarities in the configuration suggest that self-organisation of both systems may be driven by the same principles.
46. Philosopher John Searle argued that mind is more than computation. See, for example, a detailed account of Searle's Chinese Room Argument. David Cole, 'The Chinese Room Argument,' *The Stanford Encyclopedia of Philosophy* (Winter 2020 Edition), ed. Edward N. Zalta, https://plato.stanford.edu/archives/win2020/entries/chinese-room.
47. Maturana and Varela, *Autopoiesis and Cognition*, 71.
48. Maturana and Varela, *Autopoiesis and Cognition*, 73–76.
49. J. Scott Turner, 'Homeostasis as a fundamental principle for a coherent theory of brains,' *Philosophical Transactions of the Royal Society B* 374 (2019), https://doi.org/10.1098/rstb.2018.0373.
50. See, for example, https://ahlouie.com/relational-biology.
51. I am grateful to Dorion Sagan for suggesting the word 'hyperthought' in the present context.
52. A good summary of Anaxagoras' philosophy is presented in Margaret R. O'Leary, *Anaxagoras and the Origin of Panspermia Theory* (Bloomington, IN: iUniverse, Inc., 2008).
53. Predrag Slijepčević, 'Principles of information processing and natural learning in biological systems,' *Journal for the General Philosophy of Science* 52 (2021): 227–245.
54. Bateson, *Steps to an Ecology of Mind*, 483. The quote is from the essay 'Pathologies of Epistemology', pages 478–488.
55. Martin Heidegger, *The Question Concerning Technology and Other Essays*, trans. William Lovitt (New York, London: Garland Publishing Inc., 1977), 10.
56. Dorion Sagan and Lynn Margulis used the same argument in their book *What Is Life?: The Eternal Enigma* (Berkeley and Los Angeles: University of California Press, 2000).

57. Predrag Slijepčević, 'Evolutionary epistemology: Reviewing and reviving with new data the research programme for distributed biological intelligence,' *BioSystems* 163 (2018): 23–35.
58. Slijepčević, 'Evolutionary epistemology,' *BioSystems*.

Chapter 3. Pride and Prejudice

1. The evolutionary timeline of life forms is well established. The first life forms were prokaryotic microbes, bacteria and archaea, estimated to be 3.8 billion years old. The first eukaryotes, such as protists (amoebas, paramecia, etc.), were products of bacterial and archaeal mergers known as eukaryogenesis. The first eukaryotes arose at least 1.5 billion years ago. Plants and animals emerged approximately 600 million years ago. Insects most likely originated in the Silurian, 400 million years ago.
2. The oldest fossil of *Homo sapiens* is dated to approximately 300,000 years ago, according to studies published in *Nature*. See Ewen Callaway, 'Oldest *Homo sapiens* fossil claim rewrites our species' history,' *Nature* 7 (June 2017), https://doi.org/10.1038/nature.2017.22114.
3. For details, see Erik W. Grafarend, *Linear and Nonlinear Models: Fixed Effects, Random Effects, and Mixed Models* (Berlin, New York: Walter de Gruyter, 2006).
4. Michael E. Zimmerman, 'The singularity: A crucial phase in divine self-actualization?,' *Cosmos and History: The Journal of Natural and Social Philosophy* 4 (2008): 347–370.
5. A good interpretation of Aristotle's biology, including *History of Animals*, can be found in James Lennox, 'Aristotle's Biology,' *The Stanford Encyclopedia of Philosophy* (Fall 2021 Edition), ed. Edward N. Zalta, https://plato.stanford.edu/archives/fall2021/entries/aristotle-biology.
6. The notion that the unit of evolutionary selection is equivalent to the unit of mind is elaborated in the essay 'Pathologies of Epistemology,' in Bateson, *Steps to an Ecology of Mind*, 478–488.
7. Arthur Koestler, *The Ghost in the Machine* (London: Picador, 1981), 48.
8. Jakob von Uexküll, *A Foray into the Worlds of Animals and Humans*, trans. Joseph D. O'Neil (Minneapolis, MN: University of Minnesota Press, 2010), 44–46.
9. Gerhard Roth, 'Convergent evolution of complex brains and high intelligence,' *Philosophical Transactions of the Royal Society B* 370 (2015), https://doi.org/10.1098/rstb.2015.0049.
10. See, for example, David Chalmers, 'Facing up to the problem of consciousness,' *Journal of Consciousness Studies* 2 (1995): 200–219.
11. Margulis and Sagan, *What Is Life?*, 150.
12. Yinon M. Bar-On, Rob Philips and Ron Milo, 'The biomass distribution on Earth,' *Proceedings of the National Academy of Sciences, USA* 115 (2018): 6506–6511.
13. *Bioastronomy News*, 1995 (Vol. 7, Nos. 3 & 4).
14. The notion of the technological singularity, a hypothetical point in human technological development, predicted by Ray Kurzweil to occur in 2045, marks the emergence of machine superintelligence far more advanced than human intelligence.
15. Bar-On et al., 'The biomass distribution on Earth,' *Proceedings of the National Academy of Sciences, USA*.
16. Anthony Trewavas, 'The foundations of plant intelligence,' *Interface Focus* 7 (2017), https://doi.org/10.1098/rsfs.2016.0098.
17. Shapiro, 'Bacteria are small but not stupid,' *Studies in History and Philosophy of Biological and Biomedical Sciences*.
18. Ruddiman et al., 'Defining the epoch we live in,' *Science*.
19. For Earth stewardship, see, for example, F. Stuart Chapin III et al., 'Earth stewardship: Science for action to sustain the human–Earth system,' *Ecosphere* 2 (2011): Article 89. The term 'noosphere', the thinking part of the biosphere, was coined by Vladimir Vernadsky. Some proponents of Anthropocene argue that this term anticipates the technosphere behind the Anthropocene.

20. United Nations Framework Convention on Climate Change, 'The Paris Agreement,' https://unfccc.int/process-and-meetings/the-paris-agreement/the-paris-agreement.
21. Bruce Clarke, *Gaian Systems: Lynn Margulis, Neocybernetics, and the End of the Anthropocene* (Minneapolis, MN: University of Minnesota Press, 2020), 14.
22. This is an excerpt from a speech by Sir Martin Rees at the 2019 Global Challenges Summit held in London: 'Spaceship Earth is hurtling through the void. Its passengers are anxious and fractious, and their life support system is vulnerable to disruption and breakdowns. And there is too little planning, too little horizon-scanning. Our Earth has existed for forty-five million centuries, but this century is special: it's the first when one species, ours, has the planet's future in its hands.'
23. Autopoietic systems – that is, living systems, from cells to Gaia – cannot be modelled or controlled using the standard approaches of feedback control theory. For example, feedback control theory suggests that changes in atmospheric composition (e.g. an increase in CO_2 concentration) can be reversed by reducing or removing sources of CO_2 production (e.g. coal burning for electricity production). On the contrary, autopoietic systems do not rely on feedback control because they are homeorhetic. Thus, computational (cybernetic) models do not capture the dynamics of autopoietic systems. For a detailed discussion, see Sergio Rubin, Tomas Veloz and Pedro Maldonado, 'Beyond planetary-scale feedback self-regulation: Gaia as an autopoietic system,' *BioSystems* 199 (2021): 104314.
24. For differences between the cybernetic and ecological metaphors for life, see Chapter 2.
25. Rubin et al., 'Beyond planetary-scale feedback self-regulation,' *BioSystems*.
26. The term 'thought collective' was used by Ludwik Fleck in his book *Genesis and Development of a Scientific Fact* (Chicago: The University of Chicago Press, 1979). The book was originally published in German in 1935 (translated by Fred Bradley and Thaddeus J. Trenn). A thought collective is typically developed when people start exchanging ideas about a specific area of knowledge. The collective is characterised by a specific mood, leading to a thought style that incorporates a series of understandings and misunderstandings. Once the thought style becomes broad enough, the collective transforms into an esoteric circle (professionals) and exoteric circles (laymen followers). For details, see Wojciech Sady, 'Ludwik Fleck,' *The Stanford Encyclopedia of Philosophy* (Winter 2021 Edition), ed. Edward N. Zalta, https://plato.stanford.edu/archives/win2021/entries/fleck.
27. Gov.uk, 'Animals to be formally recognised as sentient beings in domestic law,' https://www.gov.uk/government/news/animals-to-be-formally-recognised-as-sentient-beings-in-domestic-law.
28. For details of ant personalities, see these two articles: Elva J. H. Robinson, Ofer Feinerman and Nigel R. Franks, 'How collective comparisons emerge without individual comparisons of the options,' *Proceedings of the Royal Society B* 281 (2014), https://doi.org/10.1098/rspb.2014.0737; Thomas A. O'Shea-Wheller et al., 'Variability in individual assessment behaviour and its implications for collective decision-making,' *Proceedings of the Royal Society B* 284 (2017), https://doi.org/10.1098/rspb.2016.2237.
29. These authors have published several papers on the topic of plant sentience and consciousness. This paper offers a good overview of the topic: Anthony Trewavas et al., 'Consciousness facilitates plant behavior,' *Trends in Plant Science* 25 (2020): 216–217.
30. In *The Power of Movements in Plants*, Charles Darwin and his son Francis proposed the 'root-brain hypothesis'. The following paper offers a good commentary on the hypothesis: František Baluška et al., 'The "root-brain" hypothesis of Charles and Francis Darwin: Revival after more than 125 years,' *Plant Signaling & Behavior* 4 (2009): 1121–1127.
31. Trewavas et al., 'Consciousness facilitates plant behavior,' *Trends in Plant Science*.
32. František Baluška and Arthur Reber, 'Sentience and consciousness in single cells: How the first minds emerged in unicellular species,' *BioEssays* 41 (2019): e1800229.

33. Baluška and Reber, 'Sentience and consciousness in single cells: How the first minds emerged in unicellular species,' *BioEssays*.
34. Steve Volk, 'Can quantum physics explain consciousness? One scientist thinks it might,' *Discover Magazine* (1 March 2018).
35. Dorion Sagan, 'Introduction: Umwelt after Uexküll'. This long essay is an introduction to the English translation of Uexküll's book *A Foray into the Worlds of Animals*, 25.
36. A good revision of Margulis' SET incorporating new phylogenetic evidence is presented in Daniel B. Mills et al., 'Eukaryogenesis and oxygen in Earth history,' *Nature Ecology and Evolution* 6 (2022): 520–532.
37. Musall et al., 'Single-trial neural dynamics,' *Nature Neuroscience*.
38. Paco Calvo, Vaidurya P. Sahi and Anthony Trewavas, 'Are plants sentient?,' *Plant, Cell & Environment* 40 (2017): 2858–2869.
39. Roger Penrose said this in an interview with Lex Fridman: https://www.youtube.com/watch?v=hXgqik6HXc0.
40. Steven J. Dick, 'Interstellar humanity,' *Futures* 32 (2000): 555–567.
41. Dick, 'Interstellar humanity,' *Futures*.
42. Milan M. Ćirković, 'Post-postbiological evolution?,' *Futures* 99 (2018): 28–35. The term 'mainstream' is used in this paper to describe the idea of post-biological evolution in the context of the future of humanity.
43. The portal *Aeon* published an essay by Beth Singler on 13 June 2017 in which she argued that AI enthusiasts have a propensity to view the power of AI in religious terms: https://aeon.co/essays/why-is-the-language-of-transhumanists-and-religion-so-similar.
44. Shapiro, 'Bacteria are small but not stupid,' *Studies in History and Philosophy of Biological and Biomedical Sciences*.
45. Richard Feynman, *The Pleasure of Finding Things Out* (New York: Basic Books, 1990), 188.
46. Gale E. Christianson, *Isaac Newton: Lives and Legacies* (Oxford: Oxford University Press, 2005).
47. Bostrom, *Superintelligence*.
48. At a lecture presented in San Diego in 2017, Sir Roger Penrose argued that understanding is not computational: https://www.youtube.com/watch?v=h_VeDKVG7e0.
49. Roger Penrose quoted Paul Dirac, who said that quantum mechanics is a provisional theory. Details in an interview with Lex Fridman: https://www.youtube.com/watch?v=hXgqik6HXc0.

Chapter 4. Civilising Force

1. Boris Pasternak, *Doctor Zhivago*, trans. Max Hayward and Manya Harari (London: Wm. Collins Sons & Co., Ltd., 1958), 223.
2. A relatively simple explanation of the Standard Model is presented by CERN: https://home.cern/science/physics/standard-model.
3. John D. Barrow, *New Theories of Everything: The Quest for Ultimate Explanation* (Oxford: Oxford University Press, 2008).
4. Lee Smolin, *The Trouble with Physics: The Rise of String Theory, the Fall of a Science and What Comes Next* (London: Penguin, 2008).
5. J. Scott Turner argued that the source of energy on early Earth, capable of propping up life as a planetary force, may not have been solar energy. Instead, thermal vents and/or radioactive deposits could have played this role initially. However, solar energy was required to drive life once it became a planetary phenomenon.
6. Turner, *Purpose and Desire*, 251.

7. Ben-Jacob et al., 'Bacterial linguistic communication,' *Trends in Microbiology*; Hiroshi Kawase, Yoji Okata and Kimiaki Ito, 'Role of huge geometric circular structures in the reproduction of a marine pufferfish,' *Scientific Reports* 3 (2013), https://doi.org/10.1038/srep02106; Eric D. Schneider and Dorion Sagan, *Into the Cool: Energy Flow, Thermodynamics, and Life* (Chicago: The University of Chicago Press, 2005).
8. The tension between science on one side and arts and humanities on the other is explained in a famous lecture by C.P. Snow, 'The Two Cultures', held at the University of Cambridge in 1959 and later published as a book.
9. Vazza and Faletti, 'The quantitative comparison between the neuronal network and the cosmic web,' *Frontiers in Physics*.
10. This was explained by Wesley Stephenson on BBC News on 4 February 2012, https://www.bbc.co.uk/news/magazine-16870579.
11. Sagan and Margulis, '"Wind at life's back"' in *Beyond Mechanism*, 205–233.
12. Kauffman, *A World Beyond Physics*, 53.
13. The transcript of a lecture presented by Hofstadter contains this quote: http://bert.stuy.edu/pbrooks/ai/resources/Analogy%20as%20the%20Core%20of%20Cognition-2.pdf.
14. Ferguson, *Civilization*, 2.
15. Predrag Slijepčević, 'Insects can teach us how to create better technologies,' *The Conversation* (7 November 2016), https://theconversation.com/insects-can-teach-us-how-to-create-better-technologies-68179.
16. Antonio Lima-de-Faria, *Biological Periodicity: Its Molecular Mechanisms and Evolutionary Implications* (Greenwich, CT.: JAI Press Inc., 1996).
17. Dan-Eric Nilsson and Nansi J. Colley, 'Comparative vision: Can bacteria really see?,' *Current Biology* 26 (2016): R369–R371.
18. František Baluška and Stefano Mancuso, 'Vision in plants via plant-specific ocelli?,' *Trends in Plant Science* 21 (2016): 727–730.
19. Richard O. Prum, *The Evolution of Beauty: How Darwin's Forgotten Theory of Mate Choice Shapes the Animal World* (New York: Doubleday, 2017).
20. Ferguson, *Civilization*, 2.

Chapter 5. Communicators

1. Nigel Goldenfeld and Carl Woese, 'Biology's next revolution,' *Nature* 445 (2007): 369.
2. The website of the British supermarket Tesco has a big cat and dog food section. These particular options have been taken from the section called 'Luxury cat food': https://www.tesco.com/groceries/en-GB/shop/pets/cat-food-and-accessories/luxury-cat-food/all.
3. Tesco's website has a section called 'Rodent and insect killer': https://www.tesco.com/groceries/en-GB/shop/household/household-essentials/rodent-and-insect-killer/all.
4. See the discussion of the concept of the holon in Chapter 3 (see 'Prejudice No. 1: Intelligence', page 49). All species have assertive tendencies; one of these is the human propensity to control the part of the environment inhabited by humanity, including pest control.
5. Pierre Teilhard de Chardin, *The Phenomenon of Man* (New York: HarperCollins, 2008).
6. Frank J. Tipler, *The Physics of Immortality: Modern Cosmology, God, and the Resurrection of the Dead* (New York: Anchor Books, 1994).
7. Kurzweil, *The Age of Intelligent Machines*.
8. For Kurzweil's prediction that the singularity – the point when machines become more intelligent than human beings – will happen by 2045, see https://futurism.com/kurzweil-claims-that-the-singularity-will-happen-by-2045. For superintelligence, see Bostrom, *Superintelligence*.

9. Kurzweil's predictions are summarised in a single document available at: https://www.kurzweilai.net/images/How-My-Predictions-Are-Faring.pdf.
10. Predrag Slijepčević and Chandra Wickramasinghe, 'Reconfiguring SETI in the microbial context: Panspermia as a solution to Fermi's paradox,' *BioSystems* (2021): 104441.
11. Carl R. Woese and George E. Fox, 'Phylogenetic structure of the prokaryotic domain: The primary kingdoms,' *Proceedings of the National Academy of Sciences, USA* 74 (1977): 5088–5090.
12. Lynn Margulis and Karlene V. Schwartz, *Five Kingdoms: An Illustrated Guide to the Phyla of Life on Earth* (New York: W. H. Freeman and Company, 1998).
13. Ferguson, *Civilization*.
14. A leading expert on quorum sensing is Professor Bonnie L. Bassler from Princeton University. An interview with her conducted by *Quanta Magazine* contains a good description of quorum sensing: https://www.quantamagazine.org/bonnie-bassler-on-talkative-bacteria-and-eavesdropping-viruses-20210308. The process of electrical signal exchange is explained in Arthur Prindle et al., 'Ion channels enable electrical communication in bacterial communities,' *Nature* 527 (2015): 59–64.
15. Trewavas, 'The foundations of plant intelligence,' *Interface Focus*.
16. Stephen Jay Gould, 'Prophet for the Earth,' *Nature* 361 (1993): 311–312.
17. I have reviewed these disciplines of biology in a recent article: Slijepčević, 'Natural intelligence and anthropic reasoning,' *Biosemiotics*.
18. The concept of bacterial language, which shows features of human languages such as syntax and semantics, was developed by Eshel Ben-Jacob. See, for example, Ben-Jacob et al., 'Bacterial linguistic communication', *Trends in Microbiology*. Similarly, the concept of plant language was developed by Jarmo K. Holopainen and James D. Blonde, 'Molecular plant volatile communication,' in *Sensing in Nature*, ed. C. Lopez-Larrea (Berlin: Landes Bioscience and Springer Science+Business Media, 2012), 17–31.
19. Vanessa Sperandio, 'Striking a balance: Inter-kingdom cell-to-cell signaling, friendship or war?,' *Trends in Immunology* 25 (2004): 505–507.
20. Jan E. Leach et al., 'Communication in the phytobiome,' *Cell* 169 (2017): 587–596.
21. Sperandio, 'Striking a balance,' *Trends in Immunology*.
22. C. Strassen, 'Bergson: Rights, instincts, visions and war,' *Philosophy Now*, 124 (2018): 10–13.
23. Leach et al., 'Communication in the phytobiome,' *Cell*.
24. Leach et al., 'Communication in the phytobiome,' *Cell*.
25. James Montoya-Lerma et al., 'Leaf-cutting ants revisited: Towards rational management and control,' *International Journal of Pest Management* 58 (2012): 225–247.
26. Montoya-Lerma et al., 'Leaf-cutting ants revisited,' *International Journal of Pest Management*.
27. Robert Rosen argued, in books and scientific papers, that living systems are not computable. See Rosen, *Life Itself*.
28. See Ben-Jacob et al., 'Bacterial linguistic communication,' *Trends in Microbiology*, and Holopainen and Blonde, 'Molecular plant volatile communication,' in *Sensing in Nature*.
29. Thomas A. Sebeok, *Signs: An Introduction to Semiotics* (Toronto: University of Toronto Press, 2001).
30. Bateson, *Mind and Nature*, 92.
31. The concept of bacterial mind was elaborated by Eshel Ben-Jacob in several papers. This is the most representative example: Ben-Jacob, 'Bacterial wisdom, Gödel's theorem,' *Physica A*.
32. A useful summary of Peirce's semiotics is provided in this entry in the *Stanford Encyclopedia of Philosophy*: Albert Atkin, 'Peirce's theory of signs,' in *The Stanford Encyclopedia of Philosophy* (Summer 2013 Edition), ed. Edward N. Zalta, https://plato.stanford.edu/archives/sum2013/entries/peirce-semiotics.

33. Uexküll, *A Foray into the Worlds of Animals and Humans*.
34. Uexküll, *A Foray into the Worlds of Animals and Humans*, 43.
35. Sebeok, *Signs*.
36. See Slijepčević, 'Natural intelligence and anthropic reasoning,' *Biosemiotics*.
37. Shapiro, 'Bacteria as multicellular organisms,' *Scientific American*. The term 'bacteriosphere' has been used in Carl R. Woese, 'Microbiology in transition,' *Proceedings of the National Academy of Sciences, USA* 91 (1994): 1601–1603; Sorin Sonea and Leo G. Mathieu, 'Evolution of the genomic systems of prokaryotes and its momentous consequences,' *International Microbiology* 4 (2001): 67–71. The term 'bacterial internet' has been used by several authors, including Thomas Sebeok, Sorin Sonea, Lynn Margulis and Dorion Sagan.
38. Luanne Hall-Stoodley, J. William Costerton and Paul Stoodley, 'Bacterial biofilms: From the natural environment to infectious diseases,' *Nature Reviews Microbiology* 2 (2004): 95–108.
39. Hall-Stoodley et al., 'Bacterial biofilms,' *Nature Reviews Microbiology*.
40. Musa H. Muhammad et al., 'Beyond risk: Bacterial biofilms and their regulating approaches,' *Frontiers in Microbiology* 11 (2020), https://doi.org/10.3389/fmicb.2020.00928.
41. Michael J. Galperin, 'A census of membrane-bound and intracellular signal transduction proteins in bacteria: Bacterial IQ, extroverts and introverts,' *BMC Microbiology* 5 (2005), https://doi.org/10.1186/1471-2180-5-35.
42. Prindle et al., 'Ion channels enable electrical communication,' *Nature*.
43. Sonea and Mathieu, 'Evolution of the genomic systems,' *International Microbiology*.
44. Moelling and Broecker, 'Viruses and evolution – viruses first?' *Frontiers in Microbiology*.
45. Elie Dolgin, 'The secret social lives of viruses,' *Nature* 570 (2019): 290–292.
46. Sonea and Mathieu, 'Evolution of the genomic systems,' *International Microbiology*.
47. Goldenfeld and Woese, 'Biology's next revolution,' *Nature*; Sonea and Mathieu, 'Evolution of the genomic systems,' *International Microbiology*.
48. Goldenfeld and Woese, 'Biology's next revolution,' *Nature*.
49. Eugen V. Koonin and Yuri I. Wolf, 'Is evolution Darwinian or/and Lamarckian?,' *Biology Direct* 4 (2009), https://doi.org/10.1186/1745-6150-4-42.
50. Koonin and Wolf, 'Is evolution Darwinian or/and Lamarckian?,' *Biology Direct*.
51. This view is expressed by Professor Michael Skinner from Washington State University, based on his research of transgenerational inheritance. A good summary of Skinner's research is presented in this online essay: https://aeon.co/essays/on-epigenetics-we-need-both-darwin-s-and-lamarck-s-theories.
52. The term 'Protoctista' is used by Lynn Margulis and Karlene V. Schwartz in their book *Five Kingdoms*. However, a more common term is 'protists'. I will use protists, except when referring to Kingdom Protoctista. The process of endosymbiosis is described in Lynn Margulis' SET, mentioned in previous chapters.
53. A useful summary of different types of symbiosis, other than endosymbiosis, is presented in the following short article: https://www.nationalgeographic.org/article/symbiosis-art-living-together.
54. Kristin Schulz-Bohm et al., 'The prey's scent – volatile organic compound mediated interactions between soil bacteria and their protist predators,' *The ISME Journal* 11 (2017): 817–820.
55. Filip Husnik, 'Bacterial and archaeal symbioses with protists,' *Current Biology* 31 (2021): R862–R877.
56. Debra A. Brock et al., 'Primitive agriculture in a social amoeba,' *Nature* 469 (2011): 393–396.
57. Arielle Woznica et al., 'Mating in the closest living relatives of animals is induced by a bacterial chondroitinase,' *Cell* 170 (2017): 1175–1183.
58. Husnik, 'Bacterial and archaeal symbioses with protists,' *Current Biology*.

59. Dorion Sagan and Lynn Margulis described 'the beast with five genomes' in an eponymous article in the *Natural History Museum* magazine in June 2001: https://www.naturalhistorymag.com/htmlsite/0601/0601_feature.html.
60. Virginia Edgcomb, 'Symbiotic magnetic motility,' *Nature Microbiology* 4 (2019): 1066–1067.
61. Asma Asghar et al., 'Developmental gene regulation by an ancient intercellular communication system in social amoebae,' *Protist* 163 (2012): 25–37.
62. Arnold A.J. Mathijssen et al., 'Collective intercellular communication through ultra-fast hydrodynamic trigger waves,' *Nature* 571 (2019): 560–564.
63. The scientist involved in the study said this in a video produced by Stanford University to explain the research to the wider audience: https://www.youtube.com/watch?v=7pgRR8G_mwA.
64. Andrew Adamatzky, 'A would-be nervous system made from a slime mold,' *Artificial Life* 21 (2015): 73–91.
65. Asghar et al., 'Developmental gene regulation,' *Protist*.
66. Gerard Manning et al., 'The protist, *Monosiga brevicollis*, has a tyrosine kinase signaling network more elaborate and diverse than found in any known metazoan,' *Proceedings of the National Academy of Sciences, USA* 105 (2008): 9674–9679.
67. Manning et al., 'The protist, *Monosiga brevicollis*,' *Proceedings of the National Academy of Sciences, USA*.
68. The eight evolutionary transitions have been described by John Maynard Smith and Eörs Szathmáry in their influential book, *The Major Transitions in Evolution* (Oxford: Oxford University Press, 1997).
69. William C. Ratcliff et al., 'Experimental evolution of multicellularity,' *Proceedings of the National Academy of Sciences, USA* 109 (2012): 1595–1600.
70. Slijepčević, 'Evolutionary epistemology,' *BioSystems*.
71. Rebecca C. Taylor, Sean P. Cullen and Seamus J. Martin, 'Apoptosis: Controlled demolition at the cellular level,' *Nature Reviews, Molecular Cell Biology* 9 (2008): 231–241.
72. Baluška and Mancuso, 'Vision in plants via plant-specific ocelli?,' *Trends in Plant Science*.
73. John S. Sparks et al., 'The covert world of fish biofluorescence: A phylogenetically widespread and phenotypically variable phenomenon,' *PLoS One* 9 (2014): e83259.
74. Sparks et al., 'The covert world of fish biofluorescence,' *PLoS One*.
75. See this website: https://www.projectceti.org.
76. Adriana De O. Fidalgo and Astrid De M.P. Kleinert, 'Reproductive biology of six Brazilian Myrtaceae: Is there a syndrome associated with buzz-pollination?,' *New Zealand Journal of Botany* 47 (2009): 355–365.
77. Margaret McFall-Ngai et al., 'Animals in a bacterial world, a new imperative for the life sciences,' *Proceedings of the National Academy of Sciences, USA* 110 (2013): 3229–3236.
78. *The New Yorker* published an article by Robert Macfarlane, 'The secrets of the Wood Wide Web', on 7 August 2016, https://www.newyorker.com/tech/annals-of-technology/the-secrets-of-the-wood-wide-web.
79. Trewavas, 'The foundations of plant intelligence,' *Interface Focus*.
80. Husnik, 'Bacterial and archaeal symbioses with protists,' *Current Biology*.

Chapter 6. Engineers

1. Uexküll, *A Foray into the Worlds of Animals and Humans*, 41.
2. A good summary of evolutionary extinction events is presented on the website of the American Museum of Natural History: https://www.amnh.org/shelf-life/six-extinctions.
3. This is a composite note in which component notes (separated by semicolons) refer to parts of the long sentence in the text (separated by semicolons) in order of appearance: Mike

Hansell, 'Houses made by protists,' *Current Biology* 21 (2011): R486–R487; See 'Eusociality and the Origin of Cities', page 73; See 'The Internet of Living Things', page 16; The ability of animals to use and create tools is described in Robert W. Shumaker, Kristina R. Walkup and Benjamin B. Beck, *Animal Tool Behavior: The Use and Manufacture of Tools by Animals* (Baltimore, MD: Johns Hopkins University Press, 2011). Another useful book on the same topic is Mike Hansell, *Built by Animals: The Natural History of Animal Architecture* (Oxford: Oxford University Press, 2007); Honeybees survey flower fields to determine the optimal strategy for nectar collection. This measuring capacity is dubbed the 'honeybee algorithm', and this natural algorithm helped engineers improve internet capacity to optimally distribute users to websites. For details of the honeybee algorithm, see Sunil Nakrani and Craig Tovey, 'From honeybees to Internet servers: biomimicry for distributed management of Internet hosting centers,' *Bioinspired Biomimicry* 2 (2007): S182–S197; Dead bacterial metropolises are stromatolites. Stromatolites are usually described as laminated rocks produced when microbial mats – multilayered microbial communities – become calcified and remain preserved for a long time. For details, see Lucas J. Stal, 'Cyanobacterial mats and stromatolites,' in *Ecology of Cyanobacteria II: Their Diversity in Space and Time*, ed. Brian A. Whiton (Dordrecht, Netherlands: Springer Science+Business Media, 2012), 65–125; Details of plant engineering in the context of biomes will be presented at the end of this chapter. For details, see 'Construction of Forest Infrastructure,' page 118; Bacteria and archaea have an immune system that protects them against viruses and other invasive genetic elements. It is called CRISPR (clustered regularly interspaced short palindromic repeats). The system is based on targeting viral DNA by reorganising bacterial/archaeal DNA. Scientists adapted the bacterial/archaeal capacity for DNA rearrangement to invent a powerful genome-editing tool that may help treat human genetic diseases. Two scientists shared a Nobel Prize for Chemistry in 2020 for this discovery. A good summary of CRISPR is presented by Eric S. Lander, 'The heroes of CRISPR,' *Cell* 164 (2016): 18–28; Corina E. Tarnita et al., 'A theoretical foundation for multi-scale regular vegetation patterns,' *Nature* 541 (2017): 398–401; Ecologists recognise that many organisms act as ecosystem engineers through modifying resources available to other species. The final product of modifications is the emergence of new habitats. For details, see Justin P. Wright and Clive G. Jones, 'The concept of organisms as ecosystem engineers ten years on: Progress, limitations, and challenges,' *BioScience* 56 (2006): 203–209.

4. All entries are from Wikipedia.
5. Julian F. Vincent et al., 'Biomimetics: Its practice and theory,' *Journal of the Royal Society Interface* 3 (2006): 471–482.
6. Rosen, *Life Itself.*
7. Soviet scientist Nikolai Kardashev speculated that humanity will be able to manipulate the cosmos through various engineering practices. He categorised these manipulations into Types 1 to 3. Type 1 is manipulation on the planetary scale. Type 2 is manipulation of the solar system. Type 3 is manipulation of the galaxy. British astronomer John Barrow suggested that current human manipulations are on the borderline between Types 1 and 2. For details, see Nikolai Kardashev, 'Transmission of information by non-terrestrial civilizations,' *Soviet Astronomy* 8 (1964): 217.
8. A large majority of universities worldwide have engineering departments. Here is a link for the Faculty of Engineering at the University of Bath: https://www.bath.ac.uk/campaigns/what-is-engineering/, which provides a basic description of engineering in the context of the modern university system.
9. Sonea and Mathieu, 'Evolution of the genomic systems,' *International Microbiology*.

10. Richard Li-Hua, 'Definitions of technology,' in *A Companion to the Philosophy of Technology*, eds. Jan K.B. Olsen, Stig A. Pedersen and Vincent F. Hendricks (Chichester: Blackwell Publishing Ltd., 2013), 18–23.
11. The science of genomics and the HGP is the subject of many textbooks and popular books. A good textbook is Tom Strachan and Andrew Read, *Human Molecular Genetics*, 4th Edition (London: Garland Science, 2011). A good popular account of the topic is Matt Ridley, *Genome: The Autobiography of a Species in 23 Chapters* (London: HarperCollins, 1999).
12. Vectors are naturally occurring circular pieces of DNA, such as viruses and plasmids, or artificially generated chromosomes, such as yeast artificial chromosomes or bacterial artificial chromosomes, which serve as vehicles for foreign DNA fragments. Thus, vectors are recombinant pieces of DNA.
13. This project was launched in 2018 with the aim of cataloguing and sequencing the genomes of all eukaryotic species within ten years. For details, see Lewin A. Harris et al., 'Earth BioGenome Project: Sequencing life for the future of life,' *Proceedings of the National Academy of Sciences, USA* 115 (2018): 4325–4333. The project has a website: https://www.earthbiogenome.org.
14. For details, see an article by Xavier B. De Ros: https://theconversation.com/scientists-are-on-a-path-to-sequencing-1-million-human-genomes-and-use-big-data-to-unlock-genetic-secrets-157210.
15. The DNA sequencing methodology was invented independently by Frederick Sanger in the UK, and Walter Gilbert and Allan Maxam in the USA. The Sanger method was used during the HGP.
16. Aarti N. Desai and A. Jere, 'Next-generation sequencing: Ready for the clinics?,' *Clinical Genetics* 81 (2012): 503–510.
17. Strachan and Read, *Human Molecular Genetics*.
18. See this editorial, 'When the White House knew how to do diplomacy,' *Nature* 582 (2020): 460.
19. A 2020 documentary film, *The Pyramids of Egypt – How and Why*, reveals the engineering techniques behind Egyptian pyramid constructions. The instruments were primitive but effective tools that allowed precise alignment and levelling, effective cutting of limestone blocks and granite, and effective repositioning of limestone blocks using ramp systems. The materials were limestone blocks collected from distant quarries and transported to building sites. Granite was used to make casing stones to cover limestone blocks. Finally, a large amount of mortar was required. The knowledge consisted of various skills, ranging from the ability to align the pyramid's sides using knowledge of astronomy (alignment with the stars), to limestone and granite precision cutting and effective transportation. The organisation of production was a process that relied on a disciplined and professional workforce, rather than forced labour. There is evidence that the workforce consisted of companies with trained personnel, division of labour and subtle organisation, including housing and sustenance. The products were pyramids as tombs for pharaohs, the monarchs of ancient Egypt. The full film is available at: https://www.youtube.com/watch?v=b7GQcAbHLG0.
20. Guy Theraulaz et al., 'The formation of spatial patterns in social insects: From simple behaviours to complex structures,' *Philosophical Transactions of the Royal Society A* 361 (2003): 1263–1282.
21. Theraulaz et al., 'The formation of spatial patterns,' *Philosophical Transactions of the Royal Society A*.
22. Guy Theraulaz, Eric Bonabeau and Jean-Louis Deneubourg, 'The origin of nest complexity in social insects,' *Complexity* 3 (1998): 15–25.
23. Francis Heylighen, 'Stigmergy as a universal coordination mechanism I: Definition and components,' *Cognitive Systems Research* 38 (2016): 4–13.
24. Theraulaz et al., 'The origin of nest complexity in social insects,' *Complexity*.

25. This phenomenon has been investigated by J. Scott Turner. Journalist Lisa Margonelli described Turner's work in an article in *National Geographic* on 2 August 2014: https://www.nationalgeographic.com/animals/article/140731-termites-mounds-insects-entomology-science.
26. Theraulaz et al., 'The origin of nest complexity in social insects,' *Complexity*.
27. H.R. Hepburn, C.W.W. Pirk and O. Duangphakdee, *Honeybee Nests: Composition, Structure, Function* (Heidelberg: Springer, 2014).
28. Bert Hölldobler and Edward O. Wilson, *Superorganism: The Beauty, Elegance, and Strangeness of Insect Societies* (New York: W. W. Norton & Company, 2009).
29. See Lisa Margonelli's description of J. Scott Turner's work: https://www.nationalgeographic.com/animals/article/140731-termites-mounds-insects-entomology-science.
30. Hölldobler and Wilson, *Superorganism*.
31. L. Keller, 'Queen lifespan and colony characteristics in ants and termites,' *Insectes Sociaux* 45 (1998): 235–246.
32. Hepburn et al., *Honeybee Nests*.
33. Shumaker et al., *Animal Tool Behavior*.
34. Heidegger, *The Question Concerning Technology*.
35. The website for the Earth BioGenome Project (https://www.earthbiogenome.org) contains a statement of 'A Grand Vision' which reads: 'Create a new foundation for biology to drive solutions for preserving biodiversity and sustaining human societies.' The website was visited on 5 October 2021.
36. Heidegger uses the term 'primal truth' on page 28 in his essay 'The Question Concerning Technology'. On page 22, he explains what he means by primal thinking: 'Therefore, in the realm of thinking, a painstaking effort to think through still more primally what was primally thought is not the absurd wish to revive what is past, but rather the sober readiness to be astounded before the coming of what is early.'
37. Heidegger, *The Question Concerning Technology*, 10.
38. Kardashev, 'Transmission of information by non-terrestrial civilizations,' *Soviet Astronomy*.
39. Koestler, *The Ghost in the Machine*, 297–298.
40. Heidegger, *The Question Concerning Technology*, 12.
41. See, for example, Shoshana Zuboff, *The Age of Surveillance Capitalism: The Fight for a Human Future at the New Frontier of Power* (London: Profile Books, 2019).
42. Maturana and Varela, *Autopoiesis and Cognition*.
43. Bateson, *Steps to an Ecology of Mind*, 478.
44. Humberto R. Maturana and Francisco J. Varela, *The Tree of Knowledge: The Biological Roots of Human Understanding* (Boston and London: Shambhala Books, 1998), 23.
45. The Intergovernmental Panel on Climate Change provides information about the anthropogenic effects on climate change in regular reports. The sixth assessment report was published in August 2021.
46. Paula Watnick and Roberto Kolter, 'Biofilms, city of microbes,' *Journal of Bacteriology* 182 (2000): 2675–2679.
47. Paula et al., 'Dynamics of bacterial population growth,' *Nature Communications*.
48. Hans-Curt Flemming et al., 'Biofilms: an emergent form of bacterial life,' *Nature Reviews Microbiology* 14 (2016): 563–575.
49. Hans-Curt Flemming and Jost Wingender, 'The biofilm matrix,' *Nature Reviews Microbiology*, 8 (2010): 623–633.
50. Timo Conradi et al., 'An operational definition of the biome for global change research,' *New Phytologist* 227 (2020): 1294–1306.

51. Zeqing Ma et al., 'Evolutionary history resolves global organization of root functional traits,' *Nature* 555 (2018): 94–97.
52. Mingzhen Lu and Lars O. Hedin, 'Global plant–symbiont organization and emergence of biogeochemical cycles resolved by evolution-based trait modelling,' *Nature Ecology and Evolution* 3 (2019): 239–250.
53. Ma et al., 'Evolutionary history resolves global organization,' *Nature*.
54. Efrat Sheffer et al., 'Biome-scale nitrogen fixation strategies selected by climatic constraints on nitrogen cycle,' *Nature Plants* 1 (2015), https://doi.org/10.1038/nplants.2015.182.
55. Sophie Chao, *Forest Peoples: Numbers across the World* (Moreton-in-Marsh, UK: Forest People Programme, 2012).
56. Ceballos et al., 'Biological annihilation via the ongoing sixth mass extinction,' *Proceedings of the National Academy of Sciences, USA*.

Chapter 7. Scientists

1. This is a composite quote. *All Life Is Problem Solving* is the title of Karl Popper's book published by Routledge in 1999. The second sentence is from Popper's book *Objective Knowledge: An Evolutionary Approach* (Oxford: Clarendon Press, 1979), 261.
2. An article entitled 'Eat the world in London' was published in *The Daily Telegraph* on 18 June 2018.
3. Charles Day, 'A physicist in the kitchen,' *Physics Today* (7 September 2010).
4. Day, 'A physicist in the kitchen,' *Physics Today*.
5. Nicholas Kurti and Hervé This-Benckhard, 'Chemistry and physics in the kitchen,' *Scientific American* (April 1994): 66–71.
6. Popper, *Objective Knowledge*, 261.
7. Audrey Dussutour, Tanya Latty and Madeleine Beekman, 'Amoeboid organism solves complex nutritional challenges,' *Proceedings of the National Academy of Sciences, USA* 107 (2010): 4607–4611.
8. Dussutour et al., 'Amoeboid organism solves complex nutritional challenges,' *Proceedings of the National Academy of Sciences, USA*.
9. John T. Bonner, 'Brainless behavior: A myxomycete chooses a balanced diet,' *Proceedings of the National Academy of Sciences, USA* 107 (2010): 5267–5268.
10. Popper explained this in the essay 'Of Clouds and Clocks', published as Chapter 6 in his book *Objective Knowledge*, 206–256.
11. Toshiyuki Nakagaki, Hiroyasu Yamada and Ágota Tóth, 'Maze-solving by an amoeboid organism,' *Nature* 407 (2000): 470.
12. Atsushi Tero et al., 'Rules for biologically inspired adaptive network design,' *Science* 327 (2010): 439–442.
13. Good summaries of evolutionary epistemology are provided by Nathalie Gontier in the *Internet Encyclopedia of Philosophy* (https://iep.utm.edu/evo-epis/) and Michael Brady and William Harms in *The Stanford Encyclopedia of Philosophy* (https://plato.stanford.edu/entries/epistemology-evolutionary/#Aca). Three principles behind evolutionary epistemology rely on the writings of evolutionary epistemologist Henry C. Plotkin, who edited an important book for the field. My paper on the subject is Slijepčević, 'Evolutionary epistemology,' *BioSystems*.
14. The three formulae in this section are from Popper's essay 'Of Clouds and Clocks', published as Chapter 6 in his book *Objective Knowledge*, 206–256.
15. An essay on the portal *Aeon* by Michael Skinner, 'Unified theory of evolution,' gives an authoritative summary on the tension between Darwinism and Lamarckism: https://aeon.co/essays/on-epigenetics-we-need-both-darwin-s-and-lamarck-s-theories.

16. Writings of a leading contemporary evolutionary epistemologist, Nathalie Gontier, take a balanced look at Darwinism and neo-Darwinism, by including views such as symbiogenesis, which are closer to Lamarckism than Darwinism. A good example is Nathalie Gontier, ed. *Reticulate Evolution* (Berlin: Springer, 2015).
17. Tero et al., 'Rules for biologically inspired adaptive network design,' *Science*.
18. Bateson, *Mind and Nature*, 92.
19. Uexküll, *A Foray into the Worlds of Animals and Humans*, 41.
20. Rosen, *Life Itself.*
21. Donald T. Campbell, 'Evolutionary epistemology,' in *The Philosophy of Karl Popper*, ed. Paul A. Schlipp (LaSalle, IL: Open Court, 1974), 413–463.
22. Ben-Jacob et al., 'Bacterial linguistic communication,' *Trends in Microbiology*.
23. The invention of Velcro is an example of biologically inspired design or biomimetics. This short online story is informative enough: http://www.SwissInfo.ch/eng/Home/Archive/How_a_Swiss_invention_hooked_the_world.html?cid=5653568.
24. Popper, *Objective Knowledge*.
25. Campbell, 'Evolutionary epistemology,' in *The Philosophy of Karl Popper*.
26. Richard C. Lewontin, 'Adaptation,' *Scientific American* 239 (1978): 212–230.
27. An excellent account of Dyson's universal symbiogenesis, in the context of biology, is presented by Nathalie Gontier in the following paper: 'Universal symbiogenesis: An alternative to universal selectionist accounts of evolution,' *Symbiosis* 44 (2007): 167–181. Dyson described his ideas behind universal symbiogenesis in the following texts: Freeman Dyson, *Infinite in All Directions* (London: Penguin Books, 1988); Freeman Dyson, 'The evolution of science,' in *Evolution: Society, Science and the Universe*, ed. Andrew C. Fabian (Cambridge, UK: Cambridge University Press, 1998), 118–135; Freeman Dyson, *Origins of Life: Revised Edition* (Cambridge, UK: Cambridge University Press, 1990).
28. Sagan, 'On the origin of mitosing cells,' *Journal of Theoretical Biology*.
29. Ruddiman et al., 'Defining the epoch we live in,' *Science*.
30. Rosen, *Life Itself.*
31. A group of fifteen scientists led by J. Craig Venter produced a bacterium with a genome created by a computer. Daniel G. Gibson et al., 'Creation of a bacterial cell controlled by a chemically synthesized genome,' *Science* 329 (2010): 52–56.
32. Turner, *Purpose and Desire*, 225–227.
33. We find the idea of consciousness being widespread in the biosphere in the writings of Gregory Bateson, in particular in his book *Mind and Nature*. See also the epigraph for Chapter 1. Philosophers call this position panpsychism. See, for example, Philip Goff, *Galileo's Error: Foundations for a New Science of Consciousness* (London: Rider, 2019). However, panpsychism remains grounded in analytical philosophy, while Bateson's thinking is much closer to a true science of life – autopoiesis.
34. Cited in Rose, *Lifelines*, 83.
35. James W. Kirchner, 'The Gaia hypothesis: Fact, theory, and wishful thinking,' *Climatic Change* 52 (2002): 391–408.
36. W. Ford Doolittle, 'Darwinizing Gaia,' *Journal of Theoretical Biology* 434 (2017): 11–19.
37. Doolittle, 'Darwinizing Gaia,' *Journal of Theoretical Biology*, 11.
38. A recent authoritative account of the concept of autopoiesis is given by Fritjof Capra and Pier Luigi Luisi in their book *The Systems View of Life*.
39. Kirchner, 'The Gaia hypothesis,' *Climatic Change*.
40. Will Steffen et al., 'Trajectories of the Earth system in the Anthropocene,' *Proceedings of the National Academy of Sciences, USA* 115 (2018): 8252–8259.

41. Rubin et al., 'Beyond planetary-scale feedback self-regulation,' *BioSystems*.
42. Rosen, *Life Itself*.
43. Montevil and Mossio, 'Biological organisation as closure of constraints,' *Journal of Theoretical Biology*.
44. Rubin et al., 'Beyond planetary-scale feedback self-regulation,' *BioSystems*.
45. See the profile of Dr W. Ford Doolittle on Wikipedia: https://en.wikipedia.org/wiki/Ford_Doolittle.
46. The movie is *The Singer, Not the Song*, directed by Roy Ward Baker in 1961, starring Dirk Bogarde. The song is the eponymous track by the Rolling Stones, released as a B-side of the single 'Get Off My Cloud'. The song is inspired by the movie.
47. Doolittle, 'Darwinizing Gaia,' *Journal of Theoretical Biology*.
48. Slijepčević, 'Serial endosymbiosis theory,' *BioSystems*.
49. Doolittle, 'Darwinizing Gaia,' *Journal of Theoretical Biology*.
50. Margulis, *Symbiotic Planet*.
51. David Jablonski, 'Lessons from the past: Evolutionary impacts of mass extinctions,' *Proceedings of the National Academy of Sciences, USA* 98 (2001): 5393–5398.
52. Peter Schulte et al., 'The Chicxulub asteroid impact and mass extinction at the Cretaceous–Paleogene boundary,' *Science* 327 (2010): 1214–1218.
53. Slijepčević, 'Natural intelligence and anthropic reasoning,' *Biosemiotics*.
54. Rubin et al., 'Beyond planetary-scale feedback self-regulation,' *BioSystems*.
55. Steffen et al., 'Trajectories of the Earth system,' *Proceedings of the National Academy of Sciences, USA*.
56. Rubin et al., 'Beyond planetary-scale feedback self-regulation,' *BioSystems*.
57. Rose, *Lifelines*, 46.
58. See this website: https://ec.europa.eu/programmes/horizon2020/en/what-horizon-2020.
59. Susan Schneegans et al., *UNESCO Science Report: The Race Against Time for Smarter Development – Executive Summary* (Paris: UNESCO Publishing: 2021).
60. Capra and Luisi, *The Systems View of Life*.

Chapter 8. Doctors

1. Julian Reiss and Rachel A. Ankeny, 'Philosophy of Medicine,' in *The Stanford Encyclopedia of Philosophy* (Spring 2022 Edition), ed. Edward N. Zalta, https://plato.stanford.edu/archives/spr2022/entries/medicine.
2. Bacteria have six different types of immune systems. These will be described later in the chapter (see 'Microbial Immunity', page 149). Furthermore, the structural coupling of the global populations of bacteria and phages (viruses that infect bacteria) means that bacterial immune systems act as barriers against epidemics and pandemics in the bacteriosphere.
3. Some biologists argue that viruses act as drivers of evolution. For example, all of evolution may be seen as virus–host coevolution. See, for example, Mart Krupovic, Valerian V. Dolja and Eugene V. Koonin, 'Origin of viruses: Primordial replicators recruiting capsids from hosts,' *Nature Reviews Microbiology* 17 (2019): 449–458.
4. Bostrom, *Superintelligence*.
5. Ruddiman et al., 'Defining the epoch we live in,' *Science*.
6. The tree of life is an old concept that has taken various forms since 1735 when Carl Linnaeus classified living forms. At present, the tree of life based on genetic tracing dominates in scientific circles. The person responsible for this is Carl Woese. Using ribosomal RNA analysis, Woese obtained a phylogenetic tree consisting of three domains: Bacteria, Archaea and Eukarya. Woese discovered the Archaea domain. The Eukarya domain contains all plants and animals.

7. Hug et al., 'A new view of the tree of life,' *Nature Microbiology*.
8. Letitia Wilkins, 'Major new microbial groups expand diversity and alter our understanding of the tree of life,' *The Molecular Ecologist* (18 March 2018), https://www.molecularecologist.com/2018/03/18/major-new-microbial-groups-expand-diversity-and-alter-our-understanding-of-the-tree-of-life.
9. Chares H. Lineweaver and Aditya Chopra, 'The biological overview effect: Our place in nature,' *Journal of Big History* 3 (2019): 109–122. The term 'dogmatic common sense' originates from Whitehead, *Essays in Science and Philosophy*, 121.
10. Carl Sagan, *Pale Blue Dot: A Vision of the Human Future in Space* (New York: Ballantine Books, 1994), 5–6.
11. Lineweaver and Chopra, 'The biological overview effect,' *Journal of Big History*, 113.
12. Georges Canguilhem, *The Normal and the Pathological*, trans. C.R. Fawcett (New York: Zone Books, 1991), 199.
13. Svetislav Basara, *Kontraendorfin* (Belgrade, Serbia: Laguna, 2020), 57. In Serbian; the translation is mine.
14. Franz Kafka, *Metamorphosis*, trans. Ian Johnston (New York: Tribeca Books, 2010), 1.
15. Jens Rolff, Paul R. Johnston and Stuart Reynolds, 'Complete metamorphosis of insects,' *Philosophical Transactions of the Royal Society B* 374 (2019), https://doi.org/10.1098/rstb.2019.0063.
16. A British biogerontologist, Aubrey de Gray, claimed that humans who will live to be 1,000 years old have already been born (see, for example, http://news.bbc.co.uk/1/hi/uk/4003063.stm). Various medical interventions, which he calls 'strategies for engineered negligible senescence', will postpone ageing. However, a historian of medicine, Nathaniel Comfort, argues that 'medical promissory notes' similar to those of de Gray are not new and are probably false: https://aeon.co/ideas/why-the-hype-around-medical-genetics-is-a-public-enemy. Francis Bacon and René Descartes claimed, similarly to de Gray, that humans will live for 1,000 years and more.
17. A Silicon Valley billionaire, Larry Ellison, invests in companies that investigate the process of ageing. According to his biographer, Mark Wilson, Larry Ellison thinks that '[d]eath is just another kind of corporate opponent that can be outfoxed': https://www.theguardian.com/business/2001/jun/24/news.theobserver.
18. The basis of Margulis' SET is the fusion of microbes. For details, see 'Symbiosis (Living Together)', page 39, and 'Prejudice No. 3: Sentience and Consciousness', page 57.
19. Margulis, *Symbiotic Planet*.
20. Moelling and Broecker, 'Viruses and evolution – viruses first,' *Frontiers in Microbiology*.
21. The struggle between symbiotic microbes that leads to full-blown endosymbiosis has been described in Slijepčević, 'Serial endosymbiosis theory,' *BioSystems*.
22. Lynn Margulis and Dorion Sagan, *Acquiring Genomes* (New York: Basic Books, 2003).
23. See, for example, Slijepčević, 'Evolutionary epistemology,' *BioSystems*.
24. James A. Shapiro, *Evolution: A View from the 21st Century. Fortified* (Chicago: Cognition Press, 2022).
25. William Martin, 'Mosaic bacterial chromosomes: A challenge en route to a tree of genomes,' *BioEssays* 21 (1999): 99–104.
26. An excellent summary of bacterial immune systems is presented in Aude Bernheim and Rotem Sorek, 'The pan-immune system of bacteria: Antiviral defence as a community resource,' *Nature Reviews Microbiology* 18 (2020): 113–119.
27. These are Werner Arber, Daniel Nathans and Hamilton O. Smith.
28. Francisco J.M. Mojica et al., 'Intervening sequences of regularly spaced prokaryotic repeats derive from foreign genetic elements,' *Journal of Molecular Evolution* 60 (2005): 174–182.

29. A brief and precise history of the research behind the development of the genome-editing methodology based on CRISPR is presented by Eric Lander in an article, 'The heroes of CRISPR', published in *Cell*. Emmanuelle Charpentier and Jennifer A. Doudna received a Nobel Prize for the discovery of CRISPR-Cas9.
30. Bernheim and Sorek, 'The pan-immune system of bacteria,' *Nature Reviews Microbiology*.
31. Shapiro, 'Bacteria are small but not stupid,' *Studies in History and Philosophy of Biological and Biomedical Sciences*.
32. Details of the justification for the award can be found on the Nobel Prize website: https://www.nobelprize.org/prizes/chemistry/2020/press-release.
33. For details of how Francisco Mojica discovered CRISPR, see the interview on Labiotech.eu: https://www.labiotech.eu/interview/francis-mojica-crispr-interview. A brief history of CRISPR by Eric Lander ('The heroes of CRISPR,' *Cell*) provides an authoritative summary of the research efforts behind CRISPR.
34. Journalist Jon Cohen summarised the patent battle in an essay published by the online edition of *Science*: https://www.science.org/content/article/latest-round-crispr-patent-battle-has-apparent-victor-fight-continues.
35. The word 'macrobe' was used by Maureen A. O'Malley and John Dupré to describe macroscopic life forms: fungi, plants and animals. See Maureen A. O'Malley and John Dupré, 'Size doesn't matter: Towards a more inclusive philosophy of biology,' *Biology & Philosophy* 22 (2007): 155–191.
36. M. Klinkowski, 'Catastrophic plant diseases,' *Annual Review of Phytopathology* 8 (1970): 37–60.
37. For details of plant immune system and various modifications of it, see Resna Nishad et al., 'Modulation of plant defense system in response to microbial interactions,' *Frontiers in Microbiology* 11 (2020), https://doi.org/10.3389/fmicb.2020.01298.
38. Margaret McFall-Ngai, 'Care for the community,' *Nature* 445 (2007): 153.
39. McFall-Ngai, 'Care for the community,' *Nature*.
40. McFall-Ngai, 'Care for the community,' *Nature*.
41. McFall-Ngai, 'Care for the community,' *Nature*.
42. Moelling and Broecker, 'Viruses and evolution – viruses first?,' *Frontiers in Microbiology*.
43. Moelling and Broecker, 'Viruses and evolution – viruses first?,' *Frontiers in Microbiology*.
44. Mezcua Martin Álvaro et al., 'The origins of zoopharmacognosy: How humans learned about self-medication from animals,' *International Journal of Applied Research* 5 (2019), 73–79.
45. Gregoire Castella, Michel Chapuisat and Philippe Christe, 'Profilaxis with resin in wood ants,' *Animal Behaviour* 75 (2008): 1591–1596.
46. Timothée Brütsch and Michel Chapuisat, 'Wood ants protect their brood with tree resin,' *Animal Behaviour* 93 (2014): 157–161.
47. Castella et al., 'Profilaxis with resin in wood ants,' *Animal Behaviour*.
48. Timothée Brütsch et al., 'Wood ants produce a potent antimicrobial agent by applying formic acid on tree-collected resin,' *Ecology and Evolution* 7 (2017): 2249–2254.
49. These papers describe the warring practices of termite-hunting ants, *Megaponera analis*, and their medical practices: Erik T. Frank et al., 'Saving the injured: Rescue behavior in the termite-hunting ant *Megaponera analis*,' *Science Advances* 3 (2017), https://doi.org/10.1126/sciadv.1602187; Erik T. Frank, Marten Wehrhahn and K. Eduard Linsemair, 'Wound treatment and selective help in a termite hunting ant,' *Proceedings of the Royal Society B*, 285 (2018), https://doi.org/10.1098/rspb.2017.2457.
50. See the above papers by Erik T. Frank et al.
51. Joel Shurkin, 'Animals that self-medicate,' *Proceedings of the National Academy of Sciences, USA* 111 (2014): 17339–17341.

52. Shurkin, 'Animals that self-medicate,' *Proceedings of the National Academy of Sciences, USA*.
53. Michael A. Huffman, 'Folklore, animal self-medication, and phytotherapy: Something old, something new, something borrowed, some things true,' *Planta Medica* 88 (2021): 187–199.
54. Huffman, 'Folklore, animal self-medication, and phytotherapy,' *Planta Medica*.
55. Michael A. Huffman, 'Self-medicative behaviour in the African great apes: An evolutionary perspective into the origins of human traditional medicine,' *BioScience* 8 (2001): 651–661.
56. Huffman, 'Self-medicative behaviour in the African great apes,' *BioScience*.
57. J.T. Gradé, John R.S. Tabuti and Patrick Van Damme, 'Four-footed pharmacists: Indications of self-medicating livestock in Karamoja, Uganda,' *Economic Botany* 63 (2008): 29–42.
58. A list of Huffman's publications is available at: https://academictree.org/primate/publications.php?pid=50553.
59. Álvaro et al., 'The origins of zoopharmacognosy,' *International Journal of Applied Research*.
60. Elizabeth A. Scott et al., 'A 21st-century view of infection control in everyday settings: Moving from the germ theory of disease to the microbial theory of health,' *American Journal of Infection Control* 48 (2020): 1387–1392.

Chapter 9. Artists

1. Richard O. Prum, 'Coevolutionary aesthetics in human and biotic artworlds,' *Biology & Philosophy* 28 (2013): 811–832.
2. Darwin, *The Descent of Man*.
3. A detailed account of the criticism of Darwin's sexual selection theory immediately after publication and later is presented by Richard O. Prum in the book *The Evolution of Beauty*.
4. Richard O. Prum calls the idea of sexual selection 'Darwin's dangerous idea'. See Chapter 1, 'Darwin's Really Dangerous Idea,' in *The Evolution of Beauty*, 17–53.
5. Prum's theory of art is outlined in Prum, 'Coevolutionary aesthetics in human and biotic artworlds,' *Biology & Philosophy*. A popular account of the theory, aimed at the general audience, is presented in Prum's book *The Evolution of Beauty*.
6. John Dewey, *Art as Experience* (New York: Capricorn Books, G. P. Putnam's Sons, 1958).
7. The first chapter of *Art as Experience* is entitled 'The Live Creature'. Dewey argues that aesthetic experience goes beyond the human world.
8. Dewey, *Art as Experience*, 15.
9. The terms 'mysterious circles' and 'underwater crop circles' were created by journalists.
10. Kawase et al., 'Role of huge geometric circular structures,' *Scientific Reports*.
11. Kawase et al., 'Role of huge geometric circular structures,' *Scientific Reports*.
12. The full video is available on the BBC website.
13. Todd Bond et al., 'Mystery pufferfish create elaborate circular nests at mesophotic depths in Australia,' *Journal of Fish Biology* 97 (2020): 1401–1407.
14. Prum, 'Coevolutionary aesthetics in human and biotic artworlds,' *Biology & Philosophy*.
15. Prum, 'Coevolutionary aesthetics in human and biotic artworlds,' *Biology & Philosophy*.
16. Prum, 'Coevolutionary aesthetics in human and biotic artworlds,' *Biology & Philosophy*.
17. Prum, 'Coevolutionary aesthetics in human and biotic artworlds,' *Biology & Philosophy*, 816.
18. Prum, 'Coevolutionary aesthetics in human and biotic artworlds,' *Biology & Philosophy*, 815.
19. Jasper Hoffmeyer, 'Introduction: Semiotic scaffolding,' *Biosemiotics* 8 (2015): 153–158.
20. Hoffmeyer, 'Introduction: Semiotic scaffolding,' *Biosemiotics*, 154.
21. Dewey, *Art as Experience*.
22. Dewey, *Art as Experience*, 15.
23. Details behind the honeybee algorithm have been presented in Chapter 6.

24. For details of the Golden Goose Award, see: https://www.goldengooseaward.org/01awardees/honey-bee-algorithm. The information on how much the honeybee algorithm contributed to saving within the internet industry was provided by Professor Thomas D. Seeley to the author (personal communication in an e-mail).
25. Dewey, *Art as Experience*, 24.
26. Dewey explains the flow of energy in nature in Chapter 1 of *Art as Experience*, 'The Live Creature'. For example, on page 14, he says: 'Order is not imposed from without but is made out of the relations of harmonious interactions that energies bear to one another.'
27. Most textbooks explain the principles behind photosynthesis using a combination of physics, chemistry and biology, reduced to the level of molecular biology. A good example is Harvey Lodish et al., *Molecular Cell Biology*, 5th Edition (New York: W. H. Freeman and Company, 2004). The principles of photosynthesis are described on pages 331–347.
28. This is the title of a popular science book: Oliver Morton, *Eating the Sun: How Plants Power the Planet* (London: Harper Perennial, 2009). The book comprehensively explains photosynthesis, which contains some artistic/lyrical elements.
29. Dewey's museum conception of art is described in Chapter 1 of *Art as Experience*, 'The Live Creature'. Here is a relevant quote from page 8: 'Most European museums are, among other things, memorials of the rise of nationalism and imperialism. Every capital must have its own museum of painting, sculpture, etc., devoted in part to exhibiting the loot gathered by its monarchs in conquest of other nations.... The growth of capitalism has been a powerful influence in the development of the museum as the proper home for works of art, and in promotion of the idea that they are apart from the common life. The nouveaux riches, who are an important by-product of the capitalist system, have felt especially bound to surround themselves with works of fine art which, being rare, are also costly. Generally speaking, the typical collector is the typical capitalist. For evidence of good standing in the realm of higher culture, he amasses paintings, statutory, and artistic bijoux, as his stocks and bonds certify to his standing in the economic world.'
30. Dewey, *Art as Experience*, 35.
31. Dewey, *Art as Experience*, 40.
32. Vasko Popa (1922–1991) was a Yugoslav poet. The short poem 'Duck' was part of the book *Bark* (1952). The English translation of Popa's poetry, including 'Duck', was first published in Vasko Popa, *Selected Poems* (London: Penguin Books, Modern European Poets, 1969), 18. Translated by Anne Pennington, with an introduction by Ted Hughes.
33. Dewey, *Art as Experience*, 39.
34. Blaise Pascal (1623–1662), French mathematician, physicist and philosopher, wrote about human nature in his philosophical work *Thoughts*. Here is the 'thinking reed' quote: 'Man is but a reed, the weakest thing in nature: but he is a thinking reed. There is no need for the whole universe to take up arms to crush him: a vapour, a drop of water is enough to kill him. But even if the universe were to crush him, man would still be nobler than his slayer, because he knows that he is dying and the advantage the universe has over him; the universe knows none of this. Thus all our dignity consists in thought.'
35. Dewey, *Art as Experience*, 15.
36. Tom Leddy and Puolakka Kalle, 'Dewey's aesthetics,' in *The Stanford Encyclopedia of Philosophy* (Fall 2021 Edition), ed. Edward N. Zalta, https://plato.stanford.edu/cgi-bin/encyclopedia/archinfo.cgi?entry=dewey-aesthetics.
37. The symbiotic relationship between animals and microbes is contained within the term 'holobiont'. All animals have accompanying microbiota consisting mostly of bacteria, but also viruses, archaea and fungi.

38. Timothy G. Dinan et al., 'Collective unconscious: How gut microbes shape human behaviour,' *Journal of Psychiatry Research* 63 (2015): 1–9.
39. Kauffman, 'Prolegomenon to patterns in evolution,' *BioSystems*.
40. All estimates (number of atoms in the universe, number of protons with 200 amino acids, etc.) are from Kauffman, 'Prolegomenon to patterns in evolution,' *BioSystems*.
41. Non-ergodicity is explained by Stuart Kauffman in simple terms in an online entry for *Edge*: https://www.edge.org/response-detail/27104.
42. Instead of the term 'holon', Kauffman uses 'Kantian wholes': 'parts exist for and by means of the whole'. Kauffman, 'Prolegomenon to patterns in evolution,' *BioSystems*.
43. Kauffman, 'Prolegomenon to patterns in evolution,' *BioSystems*.
44. Popper, *The Open Universe*.
45. Popper, *The Open Universe*, 2–3.
46. Popper, *The Open Universe*, 174.
47. Rubin et al., 'Beyond planetary-scale feedback self-regulation,' *BioSystems*.
48. This view is best summarised by Robert Rosen in his book *Life Itself*.
49. The term 'naturalised' refers to Karl Popper's rhetorical use of it. Popper, a supporter of evolutionary epistemology (see 'Evolutionary Epistemology', page 125) used an analogy that the growth of scientific knowledge is reminiscent of evolution or natural selection. I used the analogy in Chapters 7 and 8 in the context of science and medicine. Here, I am extending it to communication, engineering and aesthetics.
50. Melissa K. Nelson and Dan Shilling, eds., *Traditional Ecological Knowledge: New Directions in Sustainability and Society* (Cambridge, UK: Cambridge University Press, 2018).
51. Rubin et al., 'Beyond planetary-scale feedback self-regulation,' *BioSystems*.
52. See 'The Question Concerning Technology', page 111.
53. Popper, *The Open Universe*.
54. Hamilton O. Smith, a scientist who discovered the first restriction enzyme, shared the Nobel Prize for Medicine for the discovery. It is interesting to read the text of his Nobel lecture – a personal account of the research behind the discovery. The story contains parallels with Dewey's interpretation of 'an experience': https://www.nobelprize.org/uploads/2018/06/smith-lecture.pdf.
55. The most recent development in the long patent battle for CRISPR-Cas rights is presented in *Science*: https://www.science.org/content/article/new-crispr-patent-hearing-continues-high-stakes-legal-battle.
56. The Berne Group is a small group of European and North American doctors, formed to address the rift between current medical education and practices on one side and advances in biomedical sciences on the other. The views of the group are summarised in a special series of three articles published in Hannes G. Pauli, Kerr L. White and Ian R. McWhinney, 'Medical education, research, and scientific thinking in the 21st century,' *Education for Health*, 13 (2000): Parts One to Three, 15–25, 165–172 and 173–186.
57. Cited in S.C. Panda, 'Medicine: Science or art?,' in *What Medicine Means to Me*, eds. Ajai R. Singh and Shakuntala A. Singh (MSM, III:6, IV:1–4, 2006), 127–138.
58. Popper, *The Open Universe*, 130.

Chapter 10. Farmers

1. Edward O. Wilson, *The Social Conquest of Earth* (New York: Liveright Publishing Corporation, 2012), 76.
2. Uexküll, *A Foray into the Worlds of Animals and Humans*, 44–45.

3. Jared Diamond, *Guns, Germs, and Steel: The Fates of Human Societies* (New York: W. W. Norton & Company, 1999), 110.
4. Kenneth H. Nealson and Pamela G. Conrad, 'Life: Past, present and future,' *Philosophical Transactions of the Royal Society B* 354 (1999): 1923–1939.
5. Plant cells contain chloroplasts. These are former cyanobacteria that integrated into the eukaryotic precursors of plant cells through the process of endosymbiosis.
6. Nealson and Conrad, 'Life: Past, present and future,' *Philosophical Transactions of the Royal Society B.*
7. Steven Mithen, *The Prehistory of the Mind: A Search for the Origins of Art, Religion and Science* (London: Phoenix, 2003).
8. Urlich U. Mueller and Nicole Gerardo, 'Fungus-farming insects: Multiple origins and diverse evolutionary histories,' *Proceedings of the National Academy of Sciences, USA* 99 (2002): 15247–15249; Brian R. Silliman and Steven Y. Newell, 'Fungal farming in a snail,' *Proceedings of the National Academy of Sciences, USA* 100 (2003): 15643–15648; Brock et al., 'Primitive agriculture in a social amoeba,' *Nature.*
9. Ian Billick et al., 'Ant–aphid interactions: Are ants friends, enemies, or both?,' *Annals of the Entomological Society of America* 100 (2007): 887–892.
10. Mueller and Gerardo, 'Fungus-farming insects,' *Proceedings of the National Academy of Sciences, USA.*
11. Mithen, *The Prehistory of the Mind.*
12. The human consumption of animals reared by farming is the major cause of biodiversity loss. See Brian B. Machovina, Kenneth J. Feeley and William J. Ripple, 'Biodiversity conservation: The key is reducing meat consumption,' *Science of the Total Environment* 536 (2015): 419–431.
13. E.O. Wilson explained HIPPO in this short video: https://www.youtube.com/watch?v=Nmwr34nvMZE.
14. These words are taken from E.O. Wilson's brief talk 'Ecosystems and the Harmony of Nature': https://www.youtube.com/watch?v=Nmwr34nvMZE, 1:36–1:47.
15. Urlich G. Mueller et al., 'The evolution of agriculture in insects,' *Annual Review of Ecology, Evolution, and Systematics* 36 (2005): 564.
16. The BBC ran a story about this on 5 November 2020: https://www.bbc.co.uk/news/world-europe-54818615.
17. The Food and Agriculture Organization of the United Nations held a meeting in 2008 in Thailand dedicated to increasing insect consumption in South East Asia as a way of reducing meat consumption.
18. Yupa Hanboonsong, Tasanee Jamjanya and Patrick B. Durst, *Six-Legged Livestock: Edible Insect Farming, Collection and Marketing in Thailand* (Food and Agriculture Organization of the United Nations, Regional Office for Asia and the Pacific, 2013).
19. The following article provides an overview of the relationship between ants and aphids: Sumana Saha, Tanusri Das and Dinendra Raychaudhury, 'Myrmecophilous association between ants and aphids – an overview,' *World News of Natural Sciences* 20 (2018): 62–77.
20. Susanne DiSalvo et al., '*Burkholderia* bacteria infectiously induce the proto-farming symbiosis of *Dictyostelium* amoebae and food bacteria,' *Proceedings of the National Academy of Sciences, USA* 112 (2015): E5029–E5037.
21. Martin Pion et al., 'Bacterial farming by the fungus *Morchella crassipes*,' *Proceedings of the Royal Society B* 280 (2013), https://doi.org/10.1098/rspb.2013.2242.
22. Wilson, *The Social Conquest of Earth.*
23. For the arguments against human eusociality, see the review of E.O. Wilson's book *The Social Conquest of Earth* by Herbert Gintis, 'Clash of the Titans,' *BioScience* 62 (2012): 987–991.

24. Kevin R. Foster and Francis L.W. Ratnieks, 'A new eusocial vertebrate?,' *Trends in Ecology and Evolution* 20 (2005): 363–364.
25. Wilson, *The Social Conquest of Earth*.
26. Bert Holldobler and Edward O. Wilson, *The Leafcutter Ants* (New York, London: W. W. Norton & Company, 2011).
27. Wilson, *The Social Conquest of Earth*.
28. This information about human population growth is available here: https://ourworldindata.org/world-population-growth.
29. Predrag Slijepčević, 'Anthropocene, capitalocene, machinocene: Illusions of the instrumental reason,' *Philosophy and Society* 30 (2019): 543–570.
30. Wilson, *The Social Conquest of Earth*.
31. See this link for the full text of the letter: https://www.nybooks.com/articles/1975/11/13/against-sociobiology.
32. The formula is elaborated in this article: William D. Hamilton, 'The genetical evolution of social behaviour, I, II,' *Journal of Theoretical Biology* 7 (1964): 1–16. The formula reads $R > c/b$ (where R is the genetic relatedness of the altruist to the beneficiary; c is the cost in number of offspring equivalents; and b is the benefit in the number of offspring equivalents).
33. Martin A. Nowak, Corina E. Tarnita and Edward O. Wilson, 'The evolution of eusociality,' *Nature* 466 (2010): 1057–1062.
34. Nowak et al., 'The evolution of eusociality,' *Nature*.
35. Anna Dornhaus, Jo-Anne Holley and Nigel R. Franks, 'Larger colonies do not have more specialized workers in the ant *Temnothorax albipennis*,' *Behavioral Ecology* 20 (2009): 922–929.
36. Holldobler and Wilson, *The Leafcutter Ants*.
37. The 23 March 2011 issue of *Nature* published five letters criticising the original 2010 paper by E.O. Wilson and his collaborators, and their response.
38. The excerpt with Wilson's interview is available in full here: https://www.youtube.com/watch?v=oqb-zRCFLbU.
39. Edward O. Wilson, *On Human Nature* (Cambridge, MA: Harvard University Press, 1978).
40. Wilson, *The Social Conquest of Earth*, 187.
41. Edward O. Wilson, *The Meaning of Human Existence* (New York: Liveright Publishing Corporation, 2014), 15.
42. Wilson, *The Social Conquest of Earth*, 193.
43. Wilson, *The Social Conquest of Earth*, 191–287.
44. Mithen, *The Prehistory of the Mind*.
45. Steven Mithen, 'Did farming arise from a misapplication of social intelligence?,' *Philosophical Transactions of the Royal Society B* 362 (2007): 705–718.
46. Cited in Mithen, 'Did farming arise,' *Philosophical Transactions of the Royal Society B*.
47. Mithen, 'Did farming arise,' *Philosophical Transactions of the Royal Society B*.
48. Wilson, *The Social Conquest of Earth*, 50.

Chapter 11. How to Swim in the River of Life?

1. Margulis, *Symbiotic Planet*, 143.
2. The quote is from the movie *The Emperor's New Clothes*, based on Hans Christian Andersen's fairy tale for children: https://www.youtube.com/watch?v=z9mQoJU-6I0, 0:19–0:29.
3. 'Saint and sinner' is a description of human nature by E.O. Wilson in the context of multilevel natural selection. At one end, groups compete with groups, favouring cooperative social traits. On the other end, individuals compete with individuals, favouring selfishness. 'The opposition

between the two levels of natural selection has resulted in a chimeric genotype in each person. It renders each of us part saint and part sinner' (Wilson, *The Social Conquest of Earth*, 289).

4. Richard J. Got III, 'Implications of the Copernican principle for our future prospects,' *Nature* 363 (1993): 315–319.
5. This is known as Fermi's paradox. A recent book-length account of the paradox is presented by Milan M. Ćirković, *The Great Silence: Science and Philosophy of Fermi's Paradox* (Oxford: Oxford University Press, 2018).
6. Richard Byrne, *The Thinking Ape: Evolutionary Origins of Intelligence* (Oxford: Oxford University Press, 1995).
7. This idea was elaborated by Eshel Ben-Jacob. A detailed explanation is presented in a Google Tech Talk lecture: https://www.youtube.com/watch?v=yJpi8SnFXHs&t=1177s.
8. Milan Kundera, *The Unbearable Lightness of Being*, trans. Michael Henry Heim (New York: HarperCollins, 2009), 150.
9. The book by Gerardo Ceballos, Anne Ehrlich and Paul Ehrlich, *The Annihilation of Nature: Human Extinction of Birds and Mammals* (Baltimore, MD: John Hopkins University Press, 2015), starts with the following sentence: 'Humanity has unleashed a massive and escalating assault on all living things on this planet.' Some of the same authors published an important scientific paper with a similar message: Ceballos et al., 'Biological annihilation via the ongoing sixth mass extinction,' *Proceedings of the National Academy of Sciences, USA*.
10. Popa, *Selected Poems*. The quote is from Ted Hughes' introduction on page 10.
11. Popa, *Selected Poems*, 19.
12. Kundera, *The Unbearable Lightness of Being*, 150.
13. Stuart A. Kauffman, *Reinventing the Sacred: A New View of Science, Reason, and Religion* (New York: Basic Books, 2010).
14. This quote is from a video that Kauffman made to promote the book *Reinventing the Sacred*: https://www.dailymotion.com/video/xbpcom.
15. O'Leary, *Anaxagoras and the Origin of Panspermia Theory*.
16. O'Leary, *Anaxagoras and the Origin of Panspermia Theory*.
17. Charles S. Peirce, 'Evolutionary love,' *The Monist* 3 (1893): 176.
18. Jakob von Uexküll, *A Theory of Meaning*, trans. Joseph D. O'Neil (Minneapolis, MN: University of Minnesota Press), 195.
19. See Kauffman's video made to promote the book *Reinventing the Sacred*: https://www.dailymotion.com/video/xbpcom.
20. This is a composite note in which component notes (separated by semicolons) refer to parts of the sentence in the text in order of appearance: Pamela Lyon, 'Environmental complexity, adaptability and bacterial cognition: Godfrey-Smith's hypothesis under the microscope,' *Biology & Philosophy* 32 (2017): 443–465; Trewavas, 'The foundations of plant intelligence,' *Interface Focus*; William B. Miller Jr., Francisco J. Enguita and Ana L. Leitão, 'Non-random genome editing and natural cellular engineering in cognition-based evolution,' *Cells* 10 (2021), https://doi.org/10.3390/cells10051125; Ben-Jacob et al., 'Bacterial linguistic communication,' *Trends in Microbiology*; In recent years, scientists discovered details of communication between whales that prompted some to call this the 'language of whales'. The final result was Project CETI (the Cetacean Translation Initiative), aimed at deciphering the language of whales; Scarlett R. Howard et al., 'Numerical cognition in honeybees enables addition and subtraction,' *Science Advances* 5 (2019), https://doi.org/10.1126/sciadv.aav0961.
21. Vazza and Faletti, 'The quantitative comparison between the neuronal network and the cosmic web,' *Frontiers in Physics*.

22. Schneider and Sagan, *Into the Cool*, 93.
23. Galperin, 'A census of membrane-bound,' *BMC Microbiology*.
24. Insect individuality is elaborated in the following sources: Robinson et al., 'How collective comparisons emerge,' *Proceedings of the Royal Society B*; O'Shea-Wheller et al., 'Variability in individual assessment behaviour,' *Proceedings of the Royal Society B*; Thomas D. Seeley, *Honeybee Democracy* (Princeton, NJ: Princeton University Press, 2010).
25. Anita W. Wooley et al., 'Evidence for a collective intelligence factor in the performance of human groups,' *Science* 330 (2010): 686–688.
26. Trewavas, 'The foundations of plant intelligence,' *Interface Focus*.
27. Snait Gissis and Eva Jablonka, eds., *Transformation of Lamarckism: From Subtle Fluids to Molecular Biology* (Cambridge, MA: MIT Press, 2011).
28. Mary Rumpho et al., 'The making of a photosynthetic animal,' *Journal of Experimental Biology* 214 (2011): 303–311.
29. Slijepčević, 'Serial endosymbiosis theory,' *BioSystems*.
30. Popper, *The Open Universe*.
31. See Buchanan and Weiss, 'Things genes can't do,' *Aeon*.
32. This thesis was elaborated by Max Horkheimer and Theodor Adorno in *The Dialectic of Enlightenment* (Stanford, CA: Stanford University Press, 2002).
33. Popper, *The Open Universe*.
34. A good philosophical argument for the dogmatism of mainstream science in all epochs is presented in Fleck, *Genesis and Development of a Scientific Fact*. Fleck used the term 'thought collective' to describe the resistance of mainstream science to groundbreaking challenges that have a strong scientific basis.

Chapter 12. Equality in Diversity

1. Paul Feyerabend, *Against Method* (London: Verso, 1984), 17.
2. Antonio Lima-de-Faria, *Molecular Evolution and Organization of the Chromosome* (Amsterdam: Elsevier, 1983), 1086.
3. A defence of Feyerabend's philosophy is presented in a recent paper: Ian James Kidd, 'Why did Feyerabend defend astrology? Integrity, virtue, and the authority of science,' *Social Epistemology* 30 (2016): 464–482, https://doi.org/10.1080/02691728.2015.1031851.
4. The discovery of benzene by August Kekulé was celebrated by the German Chemical Society in 1890, twenty-five years after his famous paper. In a lecture, Kekulé said that the idea of the benzene ring structure came to him in a dream of a mythological snake, Ouroboros, eating its own tail.
5. The phrase 'anything goes' is used in Feyerabend's book *Against Method* as a synonym for diversity of ideas – the common ground for non-dogmatic sciences. In later books (*Farewell to Reason* [London: Verso, 1987]) he relied on the ideas of some classical liberals, such John Stuart Mill, to justify 'anything goes': '...a variety of views, he said [John Stuart Mill], is needed for the production of 'well-developed human beings'; and it is needed for the improvement of civilization...' (p. 33). See also a 1993 televised interview entitled 'Paul K. Feyerabend: Anything Can Go', available on YouTube: https://www.youtube.com/watch?v=EUtzWMh1fro.
6. EES has a dedicated website detailing its rationale and members: https://extendedevolutionary synthesis.com.
7. As in the case of EES, there is a dedicated website explaining the rationale of The Third Way, and biographies of all members: https://www.thethirdwayofevolution.com.
8. From The Third Way website: https://www.thethirdwayofevolution.com.

9. Denis Noble, *The Music of Life: Biology Beyond the Genome* (Oxford: Oxford University Press, 2006).
10. Shapiro, *Evolution*.
11. Ohid Jaqub, 'Serendipity: Towards a taxonomy and a theory,' *Research Policy* (2017), https://doi.org/10.1016/j.respol.2017.10.007.
12. Peter Medawar, 'Is the scientific paper a fraud?,' *Listener* 70 (1963): 377–378.
13. The expression 'becoming of the universe' is often used by Stuart Kauffman.
14. John Horgan, 'Was philosopher Paul Feyerabend really science's "worst enemy"?,' *Scientific American* (24 October 2016), https://blogs.scientificamerican.com/cross-check/was-philosopher-paul-feyerabend-really-science-s-worst-enemy.
15. The concept of Industry 4.0, a combination of AI, gene editing, nanotechnology, robotics and other new technologies that could enable smart automation, was popularised by Klaus Schwab in 'The Fourth Industrial Revolution: What It Means and How to Respond,' *Foreign Affairs* (12 December 2015).

Epilogue: A Report to an Academy

1. Edward O. Wilson, *Biophilia* (Cambridge, MA: Harvard University Press, 1984).
2. Peirce, 'Evolutionary love,' *The Monist*.

Index

About the Author

Kornelia Balul

Predrag Slijepčević is a senior lecturer in the Department of Life Sciences at Brunel University London. He is a bio-scientist interested in the philosophy of biology. In particular, Predrag investigates how biological systems, from bacteria to animals and beyond, perceive and process environmental stimuli (that is, biological information) and how this processing, which is a form of natural learning, affects the organism–environment interactions. He aims to identify those elements in the organisation of biological systems that lead to forms of natural epistemology, or biological intelligence, that might qualify those systems as cognitive agents. He has published widely in peer-reviewed journals across all areas of this book. *Biocivilisations* is his first book.